A Brief History of Chemistry

A Brief History of
Chemistry

A Curriculum Study

by
Michael Ridenour

© Copyright 2018 by Waldorf Publications

A Brief History of Chemistry: A Curriculum Study

Published by:
Waldorf Publications
Research Institute for Waldorf Education
351 Fairview Avenue, Unit 625
Hudson, NY 12534

ISBN: 978-1-943582-95-2

Please contact patrice@waldorf-research.org with feedback on this publication as well as requests for future work.

Table of Contents

Preface ix

1. Origins 1

The foundation for much of modern thought was laid in the science and philosophy of ancient Egypt and classical Greece. The chapter will compare ancient and modern and address the changes that led from one to the other. In Egypt: the land of Chem, myth and legend, the three "gifts" of Hermes Trismegistus and the doctrine of "as above so below." The times change but we share a need to understand the nature of substance with older cultures. The science of substance has a dark side: Frank Oppenheimer's concern during the making of the first atomic bomb and the blind rush to "get this thing to go." Contrasts in classical and modern thought, deductive and inductive reasoning. Greek philosophy and parallels to modern thought. Medieval alchemy and the Philosopher's Stone. Paracelsus the man and scientist. The advent of experimental medicine begins to break free of the philosophical school of science. Becher and the last fires of alchemy. Mithridates, saturated with poison. The ever puzzling mystery of fire gives rise to the theory of phlogiston.

2. Transition Theories 21

The seventeenth century: a time of turmoil, war and the fire of new Ideas lead to the birth of empirical science. Georg Stahl and phlogiston theory, the search for the evidence of things unseen. Problems with the theory. The alkali theory of opposites, shades of Empedocles. But no one knew what an acid was! The qualitative meaning of force and the efforts of the empiricists to do away with the offending word. The atom seen by Democritus. Epicurus develops an atomic theory of free will. Lucretius' poem "On the Nature of Things" popularizes the atomism of Epicurus. Chance as a precursor to probability. René Descartes espouses the mechanical model. Empirical science has a new advocate in Robert Boyle. Robert Boyle and the mechanical model. Nature abhors a vacuum. Boyle applies the new science of atoms to explain the expansion of air to fit any size container. Deism and the Cartesian partition.

3. A New Science is Born 31
The Age of Enlightenment, despotism, exploration, new ideas and an abiding faith in reason continue the transition toward a more empirical science. Joseph Priestley the dissenting minister with a quaint hobby. Smelling vapors at the brewery. Meeting with Benjamin Franklin inspires a study of electricity. Acceptance into the Royal Society. The first Club Soda. Preliminary work with the oxides of nitrogen. A fortuitous incident sparks his greatest discovery. One mouse survives . . . but Aristotle's element of air (and water) does not. Meeting house burned. Fed up with England, Priestley joins his sons in America where he continues his research. Cavendish, the misanthropic aristocrat, discovers the fiery complement to Priestley's dephlogisticated air. The chemical composition of water. The curious ways of simultaneous discovery—was something in the air?

4. Laying the Corner Stone 42
It is the best of times and it is the worst of times . . . for chemistry. Lavoisier and the foundation of a modern science. Gypsum and "fixed water" anticipate the phenomenon of "fixed air." Persistent contradictions in phlogiston theory raise fresh doubts. Priestley's visit to France and his meeting with Lavoisier. The twelve day experiment. The mystery of phlogiston is solved. Oxygen is born. The Treatise of the Elements of Chemistry defines an element and provides a new system of naming elements and compounds. Tragedy at the hands of an old enemy meets the guillotine.

5. Laying the Foundation 50
An Interlude: The French chemist and historian, Marcellin Berthelot, laments the overpowering and degrading influence of empirical science. He fears that the world has lost its moral direction; the notion of the miraculous and the supernatural have vanished like a vain mirage . . . It has become a noisy world and to help make it meaningful a selection from T. S. Eliot's poem, Four Quartets, offers an image of the Chinese jar. The stillness as a Chinese jar is still . . . moves perpetually in its stillness . . . not the stillness of the violin while the note lasts . . . and the end and the beginning were always there . . . before the beginning and after the end . . . This will become the central metaphor for the rest of the book.

The Birth of Modern Atomic Theory: John Dalton, humble Quaker schoolmaster and man of the hour, finds a world ready for the atom. The uniform composition of the atmosphere explained. Proust and the law of constant proportions. Galvani's frog legs leap onto the voltaic pile. Electricity becomes the tool of the times. Humphry Davy and the discovery of alkali and alkali earth metals. Berzelius and the electro-chemical theory of bonding. A new table of atomic weights and new nomenclature to boot! Gay Lussac and the law of volumes. Conflict between Dalton's theory and experimental results with the gases. William Prout and the quest for primordial substance. Parmenides, here we come.

6. The Search for Order 63

Enter, the nineteenth century. A shabby entry full of confusion and argument suggests much drama lies in store. A tale of some tragedy: Avogadro's hypothesis and how it applied to gases conflicts with the electro-chemical theory of bonding. The Karlsruhe conference of 1861. Cannizaro passes out Avogadro's hypothesis at the conference. Julius Lothar Meyer: the scales dropped from my eyes! Newland's law of octaves gets the Royal Society version of the Bronx Cheer. Meyer's Periodicity of the elements based on physical properties attracts attention. Mendelyev's periodic table and the successful prediction of undiscovered elements clinches the deal. Intimations of Pythagoras. The nature of chemical bonds, polar and co-valent. Electrochemistry and the ions of Arrhenias. Michael Faraday's holistic view of the electro-magnetic field offer an alternative explanation for ions. New discoveries suggest mysterious relationships between matter and energy.

7. And the Violin Plays On 79

Revolutionary finds in physics shake the complacent world of science. Romance calls. A melodious voice predicts a strange marriage. A young Polish girl, Marja Sklodowska studies physics in Paris. X-rays? The mysterious powers of uranium lead to a quest. Marja (now Marie) becomes a research assistant under the physicist Becquerel where she meets Pierre Curie. A not so strange romance. Marie and Pierre Curie get a bizarre wedding present from a reigning monarch. Pitchblende! The search for the enigmatic element that out uraniums uranium. The discovery of Polonium and Radium lead to a Nobel prize. Pierre killed in a tragic accident. William Crookes and radiant matter. J. J. Thomson and the electron. The "plum pudding" model gives the electron an English flavor. Rutherford and his pet. Radioactive decay and half lives. The proton is born. The planetary model of the atom. Chadwich and the neutron fulfill Rutherford's prophetic vision. Electron and proton— the proto hyle of the universe. Hydrogen and cosmic fire. Heisenberg compares the modern concept of energy to the fire of Heraclitus. Patterns in the Chinese jar, the end and beginning are always there.

8. Where it's All Going 101

An outside caller from inside the jar prompts a connected response. She wants to know for whom the jar turns. The goal of the future: life inside the jar. A different kind of resonance. Remembering electrons as moving aspects of a field. The fields of Faraday prepare us for the same in biology. Bones and titanium: it's the quality that makes the difference. A little word from our sponsor: breath as a Goethean archetype. Is the world catching up with Goethe? Complexity theory and recent findings in biochemistry demand a more holistic view of organic processes. A challenge to neo-Darwinian theory. Intelligent Design and its implications. Is it science? Keep smiling, you are more than your molecules. In the Goethean spirit. Is an organism a mechanism? "Morphogenetic fields" and the "organocentric" approach to a holistic (and Goethean) view. A science of qualities. A biology of cooperation—Gaia revisited. Where does this lead evolution? The story of a small worm. Chaos and emergent order. Genes don't wag the dog . . . A scientist gets a new look at corn. The Chinese jar turns again. Nature within meets nature without as play converges on chaos to create emergent order in creative thinking. Aware that old ideas die hard, a wishful good bye to Descartes.

Bibliography 133

Preface

Though the scope of the book expanded in the course of writing it to acquire a broader and more general scope, my original purpose in writing was to provide a short readable history of chemistry for my eleventh grade chemistry block. It is a long block—as close to four weeks as possible—so I had time to develop a multi-dimensional approach to chemistry. I also had a wide range of students to consider. Classes are typically made up of interested students some who are not so interested. I had to treat the subject in a way that can woo those who might think a test tube will swallow them whole if they get too close and win those who can't wait to see the whole school go up in a ball of fire. Fortunately I find chemistry to be very exciting without the great balls of fire and am particularly fond of the human side of it. And that gave the range of purpose a special focus. We are all human, at least in our better moments, I reasoned, so why not make that the key to a wide range approach to chemistry? And that is just what this little history of chemistry tries to do—get to the human side of how the science developed and show just how human it is. That way I could show how chemistry (or any other science) is a process rather than an authoritative body of knowledge that is remote from the student and indifferent to human experience. Not only that, but perhaps even more to the point, history paralleled with some period experiments can give an orderly and developmental approach to the basic laws of chemistry. And if I did it right, I told myself, I could put the historical approach to some new use in a variety of ways. Just plain chapter summaries were one option for a class with some really good writers. Or maybe paragraphs on each major player in the process. Or thematic studies, say, of how trends of thought go from Pythagorus, to the law of constant proportions. Or, I could use the text to introduce major characters and let it act as a springboard for student reports. Anyway, the idea of such an approach was exciting—it offered a way to maximize my options. And even, I told myself perhaps too fondly, the book could be a good little read in case anyone might want a short look at the history of chemistry with a few extras added. So with a grab-bag of possibilities in mind, I looked over my bookshelf and noticed some books on chemistry I hadn't gotten around to reading and put my very human intentions to work.

Bearing in mind that the book had to be a good read that made the material accessible and even "fun work" I had some definite parameters. I had to be careful not to make it too long—didn't want to turn anybody off. I knew I had to shape the whole thing so it would get to the essence of the science and of the human in an imaginative way. I also knew that Waldorf students have high expectations and can be very picky. As teachers we can perhaps take the blame for some of that; all around high expectations being the staff of life on the daily menu. And what goes around comes around. So I knew I'd better come up with something that would be a combination of cool, fiery and smooth, yes, but which also developed an approach to science that captured the enthusiasm and intensity of the chemists who spent incredibly long hours in dirty labs analyzing concoctions of dubious smoky looking fluids, and smelling it all down in reeking rooms—just to see the beautiful mysteries of an unknown world. Yes, chemists have to be a little crazy and that was fine. My students could relate to that. But the question was, did I mix it all together in the right proportions so they could sense at least the presence of those beautiful mysteries transmuting themselves into reality? That can be tricky. It requires a very deft hand and a subtle pen—far more subtle than my clumsy attempts at getting the words to wriggle right. So on the book's first try some years ago, I was a little anxious to see how the class would go for it. And they didn't bat an eye. They just took it, read it and with a little urging discussed it and treated it like it was another day's work. They seemed to think it was the most natural thing in the world that a teacher would write something like that and that was that. I didn't even get the honor of being cool. Not even 20 seconds of fiery fame. On second thought though, why should I be surprised? After all, even as a good read this translated into some good work and lots of it. But the main thing was, it seemed to be working. It went with the flow and I tried to tell myself that down deep they were enjoying it. But there were a few groans. Part of me said this was great—no pain to gain. Yeah! This is taking them up the old hill for a little exercise! Then I thought I'd better read it again and decided to change a few things here and there. I had to admit there really were some rocks in the stream, some places the needed some more clarification, better phrasing, etc. And that is how a lot of the editing went. The students would suggest this and that or ask too many times to have a point clarified and I knew I had to go over it again. I did and all was fine. In fact they kind of liked helping write it as long as they didn't have to rewrite any summaries. Fortunately this never happened. So what follows is the result of a lot of back and forth work. In many ways it is a book written for students I taught and by students who taught me.

On the other hand, the book took a direction that wasn't necessarily the direction my students wanted to see. It kept growing. With the editing came the rethinking and with the rethinking came a little more here and a little more there. And then, finally, the inevitable happened. The last chapter grew and split like a one-celled organism and a sequel was born that attempts to tell where it is all going. The organic metaphor is doubly appropriate as the verbal growth was largely inspired by recent work in the fields of biochemistry and biology. The science of complexity that has grown from discoveries in these fields has focused on how organisms function as holistic units and depend on cooperative relationships, both in their chemistry and in their nurturing environments. And this fitted right into the direction I wanted to go for the twelfth grade. Indeed, as the book grew, so did my ideas on how to use it. I began to see that the last two chapters offered a launching pad of ideas and suggested an experimental repertoire that led directly to a study of holistic systems, both as chemical and biological phenomena. This could easily be taken into the direction of certain suggestions Rudolf Steiner made on how the twelfth grade was to take up the study of chemistry in the human body. It would also allow me to move rather nicely back into the organic direction in a way that complemented our studies in the tenth grade. Just as we passed from the organic cycles of nitrogen into the chemistry of inorganic polarities via the lime cycle in the tenth year, now we were exiting the realm of the inorganic via the electric fields generated by the electrolytes of Arrhenius and Faraday. These fields would open the way to understanding holistic systems and how they generate fields of their own. Thus the book seemed to open a certain holistic pattern to the pedagogy as well. Of course this meant a lot of work and research for yours truly. A little something, I told myself, that might make my students feel like they weren't alone. And why not. Growing is what we're all about. Never grow too old to grow, I tell myself and if my students can teach me how it's done the least I can do is plant a few growing signs that tell where it all might go.

And then there's the question of the endnotes. Truth is, there was just so much that I felt I just couldn't leave out. Then I got to thinking that the notes might be useful on other levels besides direct course support. As the scope widened it seemed like the expanding material might give the book a wider appeal that would make it useful for anyone with more general interest. Or a student with a deeper interest. So as I annotated the quotes to tell who wrote what, it made sense to throw in a little more about the how and the why. Dust off the facts and add a trusty thought or two . . . and that is how the endnotes grew. But dusting off meant casting off the dusty reputation endnotes have for not being the most exciting stuff to peruse. They required their own brand of color to be in their own endnote way, a good read. And so I colored and added a little here and a little there. And the endnotes, they became, lo and behold, something of a companion volume. A little subtext on the sly. I congratulated myself: I had succeeded in sneaking in another whole little book without defeating the original short and sweet purpose. Well, almost. But at least the information is there in case I want to refer to it in class. *Or in case that student with a deeper interest needs a little more to chew on.* There really are students like that. Some students would rather be doing exciting stuff like creative writing, drama, basketball; some students get their jollies doing chemistry. They are rare and you have to take good care of them when they do come in and smell the latest best creation in the molecular line or just want to chat. And because they deserve the best (I tell myself) they get the best, endnotes and all. Then there's always some poor student who misses a day or two due to illness or to just being worn out from too much living in too short a time. For them the book (with or without the endnotes) can provide a much needed thread of continuity. The idea then is not to just provide a little something for a few lucky souls, but also a grain of special interest for those who want it, continuity for those who need it and a meaningful introduction to all. So here it is. Right here were us students can use it the best way we can.

1
ORIGINS

As BERTHELOT STATES in the introduction to *Les Origines de L'Alchimie,* ancient chemistry is derived from a book of secret knowledge given to man when certain celestial beings made amorous and successful advances to the women of earth.[1] Unlike astronomy, which has revealed some of man's aspirations to more heavenly pursuits, it seems that the origins of chemistry hail from a desire to go the other way bring heaven down to earth. Others would point out that this might also refer to how those of a higher plane became enthusiastically coupled with affairs of a lower order. But high or low, it is hard for mere mortals to interpret such things; suffice it to say that ancient chemistry was quite literally a marriage of above and below. Other accounts carry this "as above so below" theme a step further and attribute the dawn of chemistry to Hermes Trismegistus, meaning "thrice wise," who in the Egyptian pantheon of the gods is associated with the deity Thoth, the deity of healing (hence the Greek association with the god Hermes). Thoth is also known for bringing the divine fire of writing to of the Land of Chem, as ancient Egypt was known.[2] Far from meeting the same fate as his Greek counterpart, Prometheus, Thoth-Hermes was looked upon as the guiding spirit of all-healing wisdom and the preserver of ancient secrets of nature and the gods. The religious focus of Thoth-Hermes was to transform the land of Chem—"chem" meaning black, the color of the earth brought down by the Nile each year—into a path of initiation that united human consciousness with the divine consciousness of the gods. This required a lot of transformation and that was what Egyptian initiation was all about—the transformation of the dark fertile earth without into an inner path that would reunite the soul with its cosmic origin. Thanks to their gift for writing, the Egyptians described this path of initiation in hieroglyphics which we have learned to read. They told of how they looked forward to the return of the soul to the cosmic world of light—a return that reunited the soul with Osiris, the deity of light and wisdom. When Osiris was slain by the wind god Set, he retired to the heavens where he welcomed light bearing souls who transformed themselves with wisdom and truth. Isis, his ever faithful spouse, remained on earth to help the process along. Part of that process became the science of Chem, later to be known as chemistry. And that, of course, is the part that interests us here.

To be sure, this process was a rich and substantial one, full of many great works. Sacred architecture in the temples of Karnac or the pyramids of Giza (especially the Great Pyramid) preserved a wisdom that even today proves to be astounding. From studying Egyptian architecture and reading Egyptian hieroglyphics we find for example that the Egyptians must have known the radius of the earth and the distance around the equator with an accuracy that has only been equaled in the twentieth century.[3] We also know from the extensive use of the Golden Mean in sacred architecture that the Egyptians attached a great deal of importance to proportions found in nature. This combined with a desire to unite the soul with the source of divine inspiration created both a horizontal dimension and a vertical dimension to Egyptian life that some have compared to the Christian cross.[4] For the Egyptian, however, the horizontal and vertical dimensions served to unite the above and below of life that made the soul truly human where to be truly human was to also be divine. Both in myth and in practice, the credo of Egyptian science was indeed "as above, so below." It was a science of process, both inner and outer, that married the human, the natural and the divine. For the Egyptian initiate this "thrice wise" process would leave the mind with a clear focus on how the human uplifted and fulfilled nature and render the eye clairvoyant and the spirit receptive to working with a divine being or beings.

Ah, but the times, they do change . . . or seem to. Now science serves the real and the physical and the things of everyday life. So it is not too surprising that along the way from then to now the ancient science of Chem became more involved with tilling the black soil (oil?) of human progress than with cultivating relationships with nature, other people or the gods. Part of this might be because we have become our own gods. We have created a whole new world we can hold in our hands, a world that ranges from plastic cups to satellite communication systems. And between the cups in the gutter and the satellite dish, it can seem like the healing aspect of early chemistry has been forgotten. Later we will see how this will work itself out and how something of the old will be reborn in the new. As far as we are concerned now, however, it will take us were we want to go if we consider that satellite dishes and pyramids really do have something in common. It is a universal fact of life that regardless what civilization we are in we have to deal with the physical substance of life. Even the Egyptians, with all their heavenly wisdom, had to understand the nature of matter. In this light we can see how they too shared a common goal with modern chemistry. Old or new, there is an underlying assumption that the world will somehow be a better place if we can only understand the nature of substance.

But a note of caution: the black substance of science has a dark side. Recent history tells us that any naïve assumption about science being the gateway to progress can be hazardous to our health. When an openness to all phenomena unites with a sincere search for truth we *can* hope to make the world a better place through science. But science, like anything else in the wrong hands, can go awry. Dr. Mengele of Nazi Germany and his wartime atrocities with human guinea pigs is an example of the worst rank. Mengele, like thousands of his countrymen, had allowed himself to become the tool of degenerate powers. Though one can argue that these powers are beyond the control of any one individual the fact remains that individual responsibility for one's actions is a moral issue. We had a much less drastic though comparable case of the same mindset when, in the final days of WWII, the first atomic bomb was being rushed to completion at Los Alamos. Frank Oppenheimer, brother of J. Oppenheimer who directed the project, sounds a warning note regarding those frantic last hours when the race to complete the bomb was a national priority:

> Amazing how the technology tools trap one, they're so powerful. I was impressed because most of the sort of fervor for developing the bomb came as a kind of anti-Fascist fervor against Germany. But when VE Day came along, nobody slowed up one little bit. No one said, "Ah well, the main thing—it doesn't matter now." We all kept working. And it wasn't because we understood the significance against Japan. It was because the machinery had caught us in its trap and we were anxious to get this thing to go.[5]

If the story of science is the tale of man's predominance over nature, the predominance of technology over man seems to be a case of the tale wagging the dog. As in the case today with so much controversy over global warming, cloning and the misuse of genetic information (including genetically modified foods), moral questions of individual responsibility are always an issue. And many scientists warn of what can happen or will happen if we do not heed the warning signs that are ever more obvious in our human experience and in our environment. Frank Oppenheimer's alarm sounds a familiar call; the need to "get this thing to go" coupled with the need to "get the dough" are, alas, the themes of endless tragedy. Indeed, the lesson seems to be that the Egyptians had good reason to insist on science being accompanied by a more responsible moral consciousness. We will see some hopeful signs at the end of our brief history that a new turn in this direction is happening.

The French have a saying that the more things change, the more they stay the same. To some extent the history of chemistry bears this out. Though the way we perceive the world around us has completely reversed itself in the last four hundred years many of the basic assumptions and concepts remain unchanged. Yet there has been a fundamental twist that has given modern thought a new direction. Instead of looking for universal truths pertaining to love, hate, or our place in the cosmic order of things, as was the custom inherited from the Greeks, modern science has become a search for truths of the universe, truths that tell the acceleration of a falling object, the speed of light, or the distance to the nearest quasar. The Greek philosophers started from the top; they sought the reasons for existence in universal principles and then looked for how these principles worked in the details of everyday life. Modern science begins at the bottom with the details and builds them into a comprehensive view or theory. The first method is called deductive in that it deduces from a holistic principle how fundamental truths apply to the world *as a whole*. The second is called inductive as it derives from analytical detail a view of how *all the parts work together.* An example of deductive reasoning can be found in the way we apply the universal principle of breath—a principle that inhales as much as we like to exhale—to how metals expand or contract, to how plants expand upward to bloom and contract into a seed, etc. A classic example of inductive reasoning might be the boy who tears apart a clock to see how it works and then has to figure out how to get the pieces to tell time. It will become clear that one of the purposes of this course will be to show how these two approaches to knowledge, the one starting with the holistic approach, the other with the analytical, can complement each other and essentially end up at the same place. Yet the two modes of thought and their effect on individual consciousness can be very different. To see how this can be and to see how we owe so much to both modes of thought, we will often need to compare the modern with the ancient. And because we owe so much of our modern thought to the Greeks (who borrowed extensively from Egypt) we will take a brief look at those Hellenistic and holistic attitudes that laid the foundation for Western philosophy and modern science.

Berthelot quotes Parmenides, a fifth century Greek philosopher, as saying that "everything stems from a uniquely eternal and immobile essence."[6] Other axiomatic statements such as "one is all, by this all is" are common in Greek alchemical and philosophical writings. One-ness was seen to represent a divine wholeness that split or engendered two-ness when the world was created. All opposites therefore come from one-ness. But a more modern mind might do some head scratching and ask, "What on earth does this mean?" So consider a cloud. It is one and uniform—a body of water vapor—and being on the heavenly side of life, as good an example of heavenly one-ness as we are apt to find in everyday life. So we have this cloud, this earthly example of one-ness. Then raindrops form in the cloud and it is no longer one or uniform, it contains water and vapor. Aha! Two-ness. Parmenides would say the cloud represents the "uniquely eternal and immobile essence." The drop represents the first step in creation. But it is a very special drop—the very smallest drop beyond which one cannot get any smaller. Otherwise there would be nothing. This smallest "drop" is what he calls beginning or primordial substance. According to Parmenides, all matter is made of these smallest drops. Pack them together in different ways and we have true blue substance—or maybe some true yellow substance (when the sun shines). All fresh from heaven to your doorstep. But don't laugh. As we shall see, he was very close to hitting a modern mark; an important part of modern physics called particle theory is all about finding the smallest drop.

Another Greek philosopher whose thought left marks on Western science was Pythagoras. In contrast to Parmenides who sought a unifying concept with his primordial substance, Pythagoras sought to connect heaven and earth. In this he was very much a student of Egypt. Like the Egyptians, he saw a harmony of nature and the divine as essential to the well-being of humankind. And the way to achieve this harmony was with numbers—numbers that told of relationships between the human, the natural and the cosmic-divine. For Pythagoras, numbers represented certain principles like the one-ness and two-ness above and the play of earthly opposites (as in yin and yang). He also saw the dynamic presence of numbers in nature. He knew of the Golden Mean from his student days in Egypt and was familiar with how the proportions derived from the Fibonacci series live in the many forms of plants and animals. We might recall how we demonstrated this in the ninth grade with the pineapple, the way leaf stems rotate up a stalk and our own bodies. We saw how the Golden Mean relates to how organisms grow; it is a proportion of nature, of earth. The circle, on the other hand, relates to how the sun and stars rise and set; it belongs to the heavenly worlds, to the cosmos.[7] The one brings harmony and beauty to earth; the other expresses a connection with the universe. But again, numbers: the cosmic and universal value of π was found by the Egyptians and by Pythagoras in the simple fraction, 22 / 7 (22 ÷ 7 = 3.1428). They knew this was not exact: the value of 3.1416 was used in the Great Pyramid.[8] But the numbers 22 and 7 were highly symbolic. Twenty-two represents both two-ness and one-ness united (as in 2 x 11). The number seven, as everybody knows, represents the seven colors in a rainbow, the seven planets and the 7 x 4 days in a lunar month. How cosmic can you get! Small wonder that for Pythagoras the harmony of heaven and earth was a question of simple ratios. All that remained was to apply simple fractions to music and human beings could tune themselves to the heavenly spheres. To achieve such lofty aims, Western music was born. It was thanks to the Pythagoreans that we have our seven-tone scale with its major and minor ratios between notes to locate the pitch in accord with simple fractions. It works as follows. If C on a piano, for example, has a frequency of f_0, the G above it will have a frequency of $3/2\, f_0$ to create an interval of a fifth. This can be seen on the scale, C, E, D, F, G, A, B (do, re, mi, fa, sol, la, ti) as being the distance between C (do) and the fifth note, G (sol). Likewise, a fourth corresponds to a C—F interval where the frequency of F is $4/3\, f_0$. This can be seen as the distance from do to fa. A major third has a C—E ratio of 5/4 and a major sixth a C—A ratio of 5/3, etc.[9] So every time we play the lyre we have special cause to remember the original seven strings in Apollo's heavenly lyre or the fact that Pan put seven reeds in nature's flute. We also have the story of how Orpheus wooed hill and dell with his seven toned lyre. We will have reason to keep this in mind when scientists start talking in the nineteenth century about how the laws of matter conform to what, in all seriousness, is called the law of octaves.

Pythagoras was not alone in adding a vital chord to modern Western culture. Heraclitus, a sixth century philosopher and younger contemporary of Pythagoras, also sounded a recurring theme with his search for permanence in an ever changing world. His well-known saying that one cannot stand in the same river twice alludes to this. Though the river as a river is permanent and will always be there, one cannot stand in the same water twice because the river as water is always flowing downstream. But the search for permanence doesn't end there. What makes a river flow, for Heraclitus, is warmth, since once frozen the river will cease to exist as a river. Hence his conclusion that warmth was the permanent essence of the universe that kept it all flowing—like a big river. For Heraclitus, the primordial essence spoke of above must be an "ever living fire" which pervades all existence through its many aspects of transformation. We will see by and by how us moderns have a similar take on something near and dear to our culture—something we even fight wars for. Stay tuned (Pythagorean pun intended).

The old adage that no man lives on an island holds just as true for the early Greek philosophers as for the scientists who created modern chemistry. This is not just because men and women are necessarily affected by their times, rather it is more because it is human nature to reckon with the times in which we live and try to understand them. The fifth century Greek philosopher Empedocles was no exception. Writing and teaching during the time of the Peloponnesian war, he saw the world as divided between opposing forces such as love and strife. His philosophy also allowed for the four elements, earth, water, air and fire to act out their opposing natures in the play of existence. And then there is Democritus, the student of Leucippus. We will never know if a feeling of human powerlessness in the face of a warring world gave rise to the notion of a world determined by atoms, but it is curious that the first attempt at atomic theory was ripened in the mind of Democritus at this tumultuous time. Atoms offered a far more fragmented look at the world than Empedocles ever dreamed of. They became the unifying principle that Democritus developed not into a science, but into a philosophy. Like the rest of the minds of his time, he was only interested in principles, not experimental fact. But because he tried to explain purely physical phenomena with an idea, he did anticipate the modern trend of adopting a theory that would require verification in the physical world.

In all the Greeks there was an implicit faith in the power of the mind to discover the meaning of life. In this way the Greeks in general and the Greek philosophers in particular are the true sons of Odysseus, the wily protégé of Athena, the goddess of wit and wisdom. And the greatest of Greek philosophers, Socrates, and his disciple, Plato, held the mind in especially high regard. Plato, the student of Socrates, saw reason as a kind of divine fire that united the inner world of the mind with the nous or universal Mind. With Plato we have a body of thought that in many ways united much of the Greek philosophy that came before him. For example, the nous or universal Mind of Plato is closely related to the "living fire" of Heraclitus. To illustrate this we have his allegory of the cave.[10] In Plato's cave prisoners are bound with their backs to a fire that casts shadows before them. Since the only thing they see is a world of shadows, it is this which they call reality. Finally, upon breaking free, a prisoner discovers the fire that creates the shadows. In so doing he discovers the power of reason that lives in his own mind. Upon venturing outside the cave, however, the prisoner discovers the sun, the light that illumines the whole world of which the individual mind is only a part. The sun represents the universal Mind that makes it possible for the individual mind to conceive of an ideal world that differed from the real and non-ideal world of everyday life. In the light of this universal Mind we learn to see how all of nature and human existence is made of patterns (archetypes and forms) that exist in the spiritual world. These patterns can also be ideal patterns of behavior such as honesty, fidelity, etc. For Plato it was the task of the wise philosopher to guide others toward a better way "patterning" their lives with such virtuous ideals and of seeing the archetypal patterns in nature. Perhaps the greatest and most consequential gift of Plato was the manner in which he sought to guide the mind toward an understanding of a more ideal way of life. The story of the cave is a case in point. For Plato the mind would have to look upward to see the enlightening sun before human beings were ready to truly find meaning and ultimate direction in life.

In Raphael's painting, *The School of Athens,* Plato holds a copy of the *Timaeus* while pointing with his right hand to the heavens and Aristotle, holding a copy of his *Ethics*, extends his right hand horizontally toward the center of the painting. In fact it is the center since the vanishing point of the painting and Aristotle's hand occupy the same point. In this way the artist chose to illustrate a Renaissance perception of how the two philosophers differed. By showing Plato pointing to the heavens and Aristotle holding a level hand over the earth the point is made that Aristotle's philosophy, being less concerned with ideals and archetypes of the spiritual world, looks more to the earth for its definition of truth. Raphael's evaluation is as perceptive as it is suggestive. It points to Aristotle as a philosopher who was more focused on here and now relationships. For example: suppose you are building a house and your design is a very postmodern mix of English Tudor, Gothic and ranch style architecture. If you try hard you can imagine such a thing—lots of battle-ready turrets, a few gargoyles overlooking the ranchy verandas and some three story stucco held in place with those Tudor looking boards . . . And what would Plato think of your house? Well, after gazing in dubious wonder at the mix of patterns and forms he might scratch his head and wonder if you'd had a sunstroke. But not Aristotle. He'd be more concerned with the quality of the stone, the grain of the wood and whether or not it were downwind from the Parthenon or your neighbor's privy.[11] And if you were being super scientific you would probably note that almost all of his observations would be rather hard to measure. But you also notice (with no small relief) that he doesn't say one word about the gargoyles in the vestibule if they make a good hat rack. It's your house, your mind, and if you want to live in your mind's idea of a house, it's up to you. But he *would* tell you to keep a true head for functional relationships and suggest that a suit of armor in the kitchen might be out of place. To this we might add that *the real question for Aristotle has more to do with how the mind sees relationships here and now and how to determine what is true from how objects relate to one another.* In this way, we might say that Aristotle gave a new and more modern twist to the Greek search for truth. He is less concerned with philosophical generalities and indeed has a level handed way of comparing objective details. His thinking appreciates the quality of relationships in a way that is distinctly lacking in the measure all number all and weigh all thinking so characteristic of modern science. But don't sell him short for all of that—we will discover in our last chapter that a science of qualities and relationships has a modern role to play.

A major stepping-stone from Greece to the Middle Ages was the Philosopher's Stone. In many ways it was a kind of Platonic ideal that existed in the mind and in other ways it was a kind of Holy Grail of alchemy that stood at the end of a long quest. The quest was half method and half inner development and half art and if the halves didn't always add up the inner struggle did. This much recalls the Egyptian approach to science with its long initiation that changed the initiate to give him or her new faculties of insight and vision. For the alchemist this amounted to new powers of understanding at the end of long search. And human nature being what it is, alchemists varied in their approach: some went for the arts of wisdom and insight (esoteric) and some went for the method (exoteric). This made the search for the Philosopher's Stone a quest for truth and/or the right method to achieve the "Great Work." This "Great Work" could consist of knowledge, the elixir of life, healing medicines or perhaps the secret of transmuting baser stuff into something that glitters. There were recipes for the Stone that combined such wild sounding ingredients as Philosophical Gold, Sophic Mercury and Salt of the Philosophers along with purified gold, silver and quicksilver (and if it wasn't pure, the whole work could fail). This would be mixed and treated with fire in the Philosopher's Egg (also called a Hermetic Vase) where a combination of pressure and fire created the "Process of the Great Work."[12] Mark that word "process." We heard it before with the Egyptians. Now we find it alive and well with alchemy. As with Egyptian science, alchemical process was seen as an art to improve on natural process.[13] Now add to the this a dash of primordial substance—the alchemical secret spice. And consider: diamonds and other gemstones are formed in the earth from the natural process of heat and pressure. If the alchemist could duplicate this, treating the right combination of substances with the right heat and pressure, it's perfectly reasonable to suppose that with the right method and process one might shift the primordial substance around bit and change some dull and drossy substance like lead into gold or something black into something that really glitters like diamonds. And with that much prepared for the Great Work, we can get into our time travel device and skip to the twenty-first century where the search for the right method/process and the desire to improve on nature lies at the heart of another Great Work, the work of modern science. And alchemy? Today we process the right method to perform alchemical feats as a matter of routine. We create artificial gemstones from diamonds to sapphires, manipulate primordial substance in nuclear reactors to transmute elements (where plutonium is more valuable than gold, alas) and after reading how genetic engineering hopes to improve on nature one wonders if the elixir of life is hype or just another hop down the yellow brick road to the Wizard of Oz. And while we're so busy improving things, some may wonder if we need a real Philosopher's Stone to improve the quality of our thinking—and provide the wisdom to know what to do when the golden bricks don't fit!

Fortunately we don't have to worry about turning yellow bricks to gold—that's for modern science to deal with. And we aren't there yet. For now it's back to the real story of alchemy, to a historical individual who was at least in the shadow of the Philosopher's Stone. Phillipus Aureolus Theophrastus Bombast von Hohenheim (1493–1541), who avoided the tamer name of Phillip by rechristening himself Paracelsus, a name that gave the second century medical writer, Celsius, a new face. And what a face it was. A sweet smelling moniker like Phill, Aury or Theo would never have done justice to the larger than life character Paracelsus became; here was a man who not only chose a grander name but also a destiny to match. Born near Zurich into a life of medicine, he took the advice of his physician father to study metallurgy and alchemy in the mines of the Tyrol mountains. From there he went to Italy and pursued a degree at the University of Ferrara.[14] After these formal studies and much traveling to acquaint himself with folk remedies and the medicines of different cultures, we then find him in Basil where he proved his medical skills by curing a local dignitary of a chronic and debilitating disease. Thanks to this success he became Medical Officer of Health at Basel. He wasn't to stay there for long, however—he was to leave, as one author writes, in an "undignified manner."[15] This is not surprising. Not one to mull over his opinions in private, he was able to offend the academic doyens of medicine at the University of Basel in grand style. What made this all the more galling and enabled Paracelsus to practice in Basel as long as he did was the fact that he had become a household name for healing. To achieve this fame he *practiced* medicine instead of just spinning out a lot of theory. He practiced his cures and treatments both on himself and on others. He believed that if you didn't try a medicine either on yourself or see its effectiveness with your own eyes it was unproven and suspect. This experimental approach was opposed to the philosophical "medicine" passed down for the works of the Arab healers, Galen and Avicenna. In sixteenth century Europe, practical medicine such a bleeding, tooth pulling, surgery, applying herbal remedies, setting broken bones and the like was done by barbers. All too frequently their attempts to heal were guided more by local superstition than by any real knowledge of herbs or the human body. Real "doctors" mixed Galen and Avicenna with local superstition and passed it off as the philosophy and theory of medicine of the day. They taught at the universities where handling a corpse was unthought of. To touch diseased tissue was seen as undignified or worse. The medicaments they prescribed often had little to do with the ailments in question and a lot to do with the wealth of the person who was ailing. Paracelsus railed against such abuses. After publicly burning the works of Galen and Avicenna in a symbolic show of contempt he challenges the medical elite of Basel with mocking ridicule:

> If your physicians only knew that their prince Galen . . . was sticking in Hell, from whence he has sent letters to me, they would make the sign of the cross upon themselves with a fox's tail. In the same way your Avicenna sits in the vestibule of the infernal portal.
>
> Come then and listen, imposters who prevail only by the authority of your high positions. After my death, my disciples will burst forth and drag you to the light and shall expose your dirty drugs . . . [16]

To be sure, we have to remember that rhetorical invective went along with the times. Luther called the obese King Henry VIII that "fat hog on the throne of England." And Paracelsus, like Luther, was a man to put deeds behind his words. An apocryphal but likely story has it that he invited the medical worthies of Basel to a lecture. These very dignified men entered a room to behold a closed vessel on the lecture table. Paracelsus began the lecture by removing the lid and revealing a huge pile of human feces. With his usual mocking style he railed against the hypocrisy of his audience for assuming to know of the body while refusing to consider the most basic of alimentary functions. He went on to extol the virtues of fermentation and the need to enliven the soil with excrement, human or otherwise. If the episode really happened there is no doubt that such a demonstration must have made a lasting impression on his listeners. But true in fact or not, it seems more than true to character; the most improbable part of the story is that it took his foes two years to get him chased out of Basel!

Part of Paracelsus' fame grew from the fact that he lectured in German instead of Latin, as was the custom of the day. Part of it was because he achieved very real cures to back up his words, cures that were notably lacking in the medical pretensions of his contemporaries. And part of it was because he breathed new life into the alchemical fires that still burned beneath the science of his time. He taught that sickness was due to an imbalance of head, blood (heart) and limbs. He referred to these parts of the body as the salt (head), mercury (heart/blood) and sulfur (metabolism/limbs) poles. And what, we might ask, is so heady about salt? Well, potato chips came after the time of Paracelsus so he didn't know how they put a little salt on the brain and wake it up. But long before Paracelsus the world knew of the stimulating power of salt. And mercury? It's easy to see how flowing mercury relates to heart and blood. That leaves fiery smelly sulfur and metabolism. Don't have to eat garlic—which contains a lot of sulfur—to know how they connect! So these general alchemical principles were not without a valid and empirical relation to how the body functions. Then he leaves the strictly empirical and turns the body into a universal principle that needed balance and harmony and experimented with medicines that gave it. In this he was a child of the Greeks and Egyptians. Yet he also foresaw a more modern approach to medicine that gave attention to the individual patient. Though his ideas of how to treat an illness were steeped in alchemical generalities, his threefold structure of head, heart and limbs, of thought, feeling and will, anticipated the modern attempt to link mental and emotional factors to our physical condition. His attempt to apply science to actual phenomena set him apart from his contemporaries and made him a much more modern figure. Yet he believed very strongly that there were unifying principles in the body and the cosmos that needed to be respected in order for a person to be in good health. For Paracelsus the body was the laboratory that proved the truth of his science. In this much he was modern. But he saw his body soul and mind as the instrument of his investigation. Through years of trial and effort he trained himself to see the cause of sickness rather than rely on, say, a stethoscope or a thermometer, neither of which had been invented in his day. In this he was a member of the old school. In many ways he was both a precursor to the modern and in many ways one of the last voices of the ancient schools of initiation that trained the soul and mind in higher powers of perception. It is significant that he died 23 years before another man of science, Galileo Galilei. Both men were very much a part of the Renaissance that was sweeping Western civilization and harbingers of ideas that were to change the way we think.

Although born almost 100 years after Paracelsus died, Johann Becher (1635–82?) fulfilled Paracelsus' prophecy that those who followed him would carry on the Great Work of alchemy and bring it to the light of day. Every bit the romantic revolutionary of Paracelsian fame, Becher is proud to carry on the alchemical tradition. In his most important book, *Physica Subterranea*, he describes himself as

> ... one to whom neither a gorgeous home, nor security of occupation, nor fame, nor health appeals to me; for me rather my chemicals amid the smoke, soot and flame of coals blown by bellows. Stronger than Hercules, I work forever in an Augean stable, blind almost from the furnace glare, my breathing (sic) affected by the vapour of mercury. I am another Mithridates saturated with poison. Deprived of the esteem and company of others, a beggar in things material, in things of the mind I am Croesus. Yet among all these evils I seem to live so happily that I would die rather than change places with a Persian king.[17]

Like Paracelsus, Becher found in alchemy an expression of world and cosmic order that connected him with the human and the divine. He saw Nature as being created by God, the ultimate chemist, and in this creation the cycles of change and exchange were set in motion to offer the world of man ample opportunity to be fruitful and multiply in mind, body and soul. He found in the mercantile community an expression of change and exchange that paralleled that of the original creation and felt that it was the role of the chemist to unveil the secret processes of God's ever-changing universe. Hence the divine responsibility of the chemist to master the ancient challenge of finding economically important minerals, not the least of which was gold. In addition to founding a technical school in Austria under the auspices of the Austrian emperor,

Leopold I, he also launched a scheme in the Netherlands for recovering gold from silver by means of sea sand. There were some ethical questions about this latter venture since Becher left the country before producing the promised gold, taking the funds of those who invested in the enterprise with him. He insisted on his sincerity, however, and returned some years later in the attempt to make good on his intentions. A second failure resulted in yet another alchemist being chased out of town. He traveled extensively after this and is supposed to have died after investigating the tin mines of Cornwall in the effort to improve on the mining techniques that had not changed all that much since the times of the Roman conquest of England in the second century CE.

Half mystic, half scientist like Paracelsus before him, Becher was more than just enamored by the fire and smoke of the chemist: the secret of fire was for him and others of the period the secret of creation just as the breath of the bellows was analogous to the breath God blew into the nostrils of Adam to raise the first man from the dust. It was paramount therefore to solve the mystery of fire before the secrets of creation could be properly understood. To this end, Becher proposed that the fiery nature of sulfur was present in all combustible materials and was released when a substance burned. All that remained was to give this essence of fire a name which Georg Stahl, a contemporary of Becher, did by calling it phlogiston. The science of chemistry was born with the analytical attempts to discover the mysteries of this elusive phlogiston, a quest that was not to end till the beginning of the nineteenth century when Lavoisier finally put the whole matter to the test with his famous balance.

Notes to Chapter 1

1. *Les Origines de L'Alchemie*, p. 1
2. The gift of writing enabled people to record events and thoughts for future generations. In this way it provided for continuity between past and present at a time when people were forgetting the old oral traditions that relied on memory (like Homer) to recall an epic past. Hence the need for writing that could recall words that spoke of immortal truths that live for ever, words that can take one prepared to read them into realms that would otherwise be inaccessible. That is one reason why writing was looked upon as divinely inspired.

The power of words is the power to transform. And for Thoth/Hermes the power to transform lived in how we use words and ideas to better understand how to heal illnesses of the present and to transform the mind to better receive divine truths from the past. In this way we become better prepared to transform the earth and better provide for the future. This transformation of the earth became the science of Chem, later known as alchemy, a corruption of the Arabic al chem., and finally as modern chemistry.

3. *Secrets of the Great Pyramid*, pp. 189–213.

4. The "as above so below" theme takes on a literal meaning in the way the Egyptians considered the Milky Way to be the Nile of the Sky (*The Orion Mystery*, pp. 119–122). But the connection between heaven and earth is a universal theme in the mythology that also affects our own culture. In the Judo-Christian tradition, which shows many similarities with the Egyptian, we have in Genesis how Adam was raised from the dust by divine breath and how he was created in God's image. In Plutarch's *Moralia*, vol. V, we also have the story of Isis and Osiris that allows Osiris a place in the stars to welcome worthy souls. The whole story of Isis and Osiris with the death of Osiris and Isis' birth of a redeemer son, Horus, contains many parallels to Christian symbolism. Christian Jacq quotes from the Egyptian *Sarcophagus Texts* where the divine voice says, "I am eternal, I am the Light . . . I am the one who created the Word, I am the Word. . ." (*Le Message Initiatique Des Cathédrals*, p. 79). This recalls the beginning of the Gospel of John: "In the beginning was the Word, and the Word was with God, and the Word was God . . ." There are many other parallels that didn't miss the early Christians. St. Augustine was to remark upon considering the mummies and burial rites of the Egyptians that they were the only Christians who truly believed in the resurrection (Ibid, p. 83). The common cause between Egypt and Christianity was not lost to the esoteric school of alchemy either. It is very likely that the esoteric side of alchemy with its emphasis on how alchemy is as much an inward transformation as an outer one owed its science to Egyptian sources.

5. *The Double-Edged Helix, Genetic Engineering in the Real World*, p. 4.

6. *Les Origines de L'Alchimie*, p. 8

7. The 3.1416 value of π is given by 6/5 of the Golden Mean, φ, squared or $\pi = 6/5\varphi^2$. This provided yet another confirmation of the fundamental relationship between heaven and earth.

8. The Great Pyramid literally conceals a whole world of secrets that link π with the value of the Golden Mean, φ. The Pyramid is built to duplicate the dimension of the northern hemisphere with the height corresponding to the radius of the earth at the north pole (and slightly shorter than the radius at the equator) and the sides corresponding to the latitude and longitude of the Pyramid's location. (See *Secrets of the Great Pyramid*, pp. 189–216.)

9. *The Science of Musical Sound*, p. 64

10. This story, though found at the beginning of Book Seven in Plato's Republic, is really of Pythagorean origin and as such is probably Egyptian. (See *Le Nombre d'Or*, Vol. II, p. 9.)

11. These observations roughly coincide with Aristotle's nine or ten categories. The list usually includes quantity, substance, quality, relation, position, time, position, action and passivity. In *Aristotle*, pp. 21–23, Sir David Ross states that Aristotle takes no pains to be consistent over the exact number of them.

12. The relation between Egypt and medieval European society remained a close one notwithstanding the burning of the library of Alexandria. Christian Jacq writes that these links were nurtured by commerce and the exchange of art, especially by way of Byzantium. Coptic and Byzantine influence spread to adorn the altars of German religious orders with ivory carvings and other treasures. Between the sixth and eighth centuries Merovingian scribes, especially Grégoire de Tours, often referred to oriental merchants who were established in several French cities with the collective name of "Syrians." That these close ties also fanned the flames of alchemy with esoteric inspiration (or exoteric speculation) would be expected considering that alchemists were among the most well traveled and cosmopolitan of artisans. (See *Le Message Initiatique des Cathédrals*, p. 90.)

13. For the alchemist as artist, see *The Norton History of Chemistry*, p. 22.

14. Some doubt has been cast as to whether or not Paracelsus actually received a degree. No record of it seems to exist. But it would have been out of character anyway—and his antagonistic feeling for academic medicine may in fact date from this period.

15. *The Norton History of Chemistry*, pp. 43–44

16. The sixteenth century was a fiery time when people used fiery words and often resorted to fiery deeds. When religious matters were at stake, charges of heresy and witchcraft all too often became hotter than words. This made Paracelsus and his outrageous raving seem all the more courageous. But behind this side of the man there was a fervent devotion also worthy of his time. I take the liberty here to quote at length from some of his own words (Parecelsus, *Selected Writings*, Bollingen Series XXVIII) to show how he was at once typical of his age and on many matters far beyond it. His practical idealism regarding women, for example, avoided the chivalric pedestal or the opposite extreme of outright misogyny so common at his time. Some of his ideas are amusing (no red bearded surgeons allowed!) and some typical for the age (his views on the mating game) but his ideas on medicine were quite advanced. His religious views, though colored by the Reformation theology of Luther, owe some of their deep regard for nature and man alike to Neo-Platonist concepts most likely inherited from his exposure to ideas of the

Italian Renaissance while in Italy.

On medicine:

> The physician does not learn everything he must know and master at high colleges alone; from time to time he must consult old women, gypsies, magicians, wayfarers, and all manner of peasant folk and random people, and learn from them; for these have more knowledge about such things than all the high colleges. (p. 57)
>
> Medicine should be taught so cleanly and clearly in the language of the homeland that the German should understand the Arab, and the Greek the German . . . (pp. 62–63)
>
> Every physician must be rich in knowledge, and not only of that which is written in books; his patients should be his book, they will never mislead him . . . and by them he will never be deceived. But he who is content with mere letters is like a dead man; and he is like dead physician. As a man and as a physician, he kills the patient. Not even a dog killer can learn his trade from books, but only from experience. And how much more is this true of the physician? (p. 50)

Some of the qualifications of a good surgeon are as follows (pp. 52–55):

A clear conscience

A gentle heart and a cheerful spirit

Moral manner of life and sobriety in all things

Greater regard for his honour than for money

Greater interest in being useful to his patient than to himself

He must not be married to a bigot

He should not be a runaway monk

He should not practice self-abuse

He must not have a red beard

He must despise no one

Regarding knowledge of the body:

He should know all the bones of the body

He should know all the blood vessels

He should know the veins and arteries of the whole body

He should know what injury can befall each organ

Regarding the practice of his art:

He should know all the vulnerary herbs

He should know all tissue-forming remedies He should know the effect

> of each remedy
>
> He should know plaster for wounds
>
> He should know lotions for wounds

On human dignity:

> Thoughts create a new heaven, a new firmament, a new source of energy, from which new arts flow . . . For such is the immensity of man that he is greater than heaven and earth. (p. 45)
>
> We are born to be awake, not to be asleep!
>
> Therefore, man, learn, and learn, question and question, and do not be ashamed of it; for only thus can you earn a name that will resound in all countries and never be forgotten. (p. 105)

On medicine as a divine calling:

> It is the physician who reveals to us the diverse miraculous works of God. And having revealed them he must use them in the right way, not in the wrong way . . . so that many people may be able to see the works of God and recognize how they can be used to cure disease. (p. 67)

On man and woman:

> A woman is like a tree bearing fruit. And man is like the fruit that the tree bears . . . (p. 26)
>
> God does not want man or woman to be like a tree which always grows the same fruit. (p. 32)
>
> God endowed man with reason, in order that he might know what desire means. But he himself must decide whether to yield to it or not, whether to let act on him or not, whether to follow his intelligence or not . . . But all this takes place only if he himself wants it; otherwise there is no seed in him . . . It is the same with woman. When she sees a man, he becomes her object, and her imagination begins to dwell on him. She does this by virtue of the ability that God has bestowed upon her . . . It is in her power to feel desire or not. If she yields, she becomes rich in seeds; if not, she has neither seed nor urge. Thus God left the seed to the free decision of man (and woman), and the decision depends upon man's will. (p. 33)

17. *The Norton History of Chemistry*, p. 79.

2
Transition Theories

Phlogiston was a child of the seventeenth century, a time of turmoil when new ideas ran like fire in the blood of Europe. The Reformation had born fruit a century before and the tide of independent thinking it brought carried with it a new desire for freedom and discovery. Even the Thirty Years War could not quell the voices of freedom that made Frankfort a refuge for free thinkers during the devastating first half of the century. The "Pilgrim fathers" (and mothers) set sail for the New World. The last of the French Protestants, the Huguenots, flee the port city La Rochelle on the western coast of France and seek refuge from England to South Africa. England survived the beheading of Charles I at the hands of a Puritan Parliament led by Cromwell. This short-lived experience in theocracy gave way to the colorful and promiscuous reign of Charles II while Reformation Part II raged on the continent to feed the every hungry stake. Christian Humanism, inspired and formed by the sixteenth century freethinker Erasmus of Rotterdam, met the fatalism of Spinoza and the materialism of Hobbes. Rembrandt applied new concepts of light and dark to painting, Galileo focused his telescope on sun spots and the moons of Jupiter and Copernicus conceived of a heliocentric solar system. Descartes philosophizes on his dictum, I think, therefore I am. But he was hardly alone; it seemed like the whole busy world was thinking. It was a time when old ideas were being reborn with new faces. It was a time when new ideas were plunging into uncharted seas of exploration. While a whole New World beckoned to explorer and revolutionary alike, a whole new horizon hailed a new age of thought. It was a time when science had to happen.

Not only did science become a major player in all of this drama, but with the fiery quality of the age, it seems almost too on cue that phlogiston take center stage. But it did and the mystery of fire (like the mystery of life itself discussed today in terms of genes and intelligent design[1]) was approached both empirically with facts in hand and philosophically with handy ideas. Georg Stahl, as a professor of medicine with a knowledge of metallurgy, was able to provide the facts in hand from what he knew of metal refining and the handy ideas from his vitalist views that life was imbued with a life force that made it separate from the mineral world. Both of these backgrounds came together to help form his ideas on phlogiston, a substance that on one hand asserted physical properties and on the other seemed to possess non-physical if not vitalistic attributes. Enough practical chemistry had been done to determine that metals could not be broken down any further by fire or other means. If attacked by acid they turned into lusterless and friable something else called a salt. But metals were shiny and thus retained a fiery nature, a nature that indicated the presence of phlogiston. Because no one had ever isolated phlogiston or identified it in the lab, its presence was theoretical.

In short, phlogiston was a handy idea. But as handy ideas go, it seemed to explain some basic phenomena. A look at a common example from metal refining will explain why. Stahl knew that metal ores were heated with carbon to obtain the pure metal. He reasoned that because metals represented a phlogiston rich element, an ore must be a phlogiston poor substance that need added phlogiston to reveal the pure metal hidden within. This phlogiston rich substance was the carbon:

ore (phlogiston poor) + carbon (phlogiston rich) = metal/phlogiston

If this was true, the reverse must also be the case. When a metal is heated, it will lose its phlogiston and become a calx or metal ore:

metal/phlogiston = calx or ore (phlogiston poor) + phlogiston (driven off in the form of extra heat)

Many things remained to be worked out. Sulfur, for example was seen at this time as a "mixt" of phlogiston and "acid former" because when burned the gas given off would form an acid when mixed with water.² But what was sulfur, really? It seemed like a veritable chameleon in the many forms it could take. Other questions begged further analysis and it was felt that further investigation would provide the needed answers. The chase for science was on. Looking back on it now, we can see that the quest for the enigmatic phlogiston was especially significant because it looked at nature as the source of truth instead of philosophical argument. The elusive explanation for phlogiston was sought not in philosophy books, not in arcane teachings of bygone civilizations, but in nature herself. That is why the search for the cause of fire, for the mysterious substance called phlogiston, opened many doors. It was a search that insisted on looking for facts "out there" as opposed to inward speculation. Most of all, phlogiston was an idea that required verification. It was an idea that required facts to support it, facts gleaned from observation. The quest for phlogiston became a questioning of substances and in doing so it led to a much clearer distinction between elements and compounds and between physical properties (weight, color, density, etc.) and chemical properties (what combines with what and how). This created a new vision of nature where theory and experiment worked hand in hand.

As the facts started coming home to roost, the new theory was put to the test. So, we might ask, did Becher and Stahl get to share some early equivalent of a Nobel Prize? Not quite. To see why we need to look more closely at how theory and fact fit or didn't fit. The theory states in various ways (which kept changing to fit the facts) that certain substances such a metals, carbon, wood, sulfur and jewels such as diamonds or sapphires were rich in phlogiston. All of these substances will either support combustion, or, as with the metals and precious stones, shine with a "fiery" luster. According to the theory, those substances that were phlogiston rich could be induced to give their phlogiston to substances that were phlogiston poor. Carbon, as we saw above, when heated with a metal ore would lose its phlogiston and thereby cause the ore to turn into a pure metal. Since phlogiston was supposed to be substantial, that is, have weight, the pure metal should therefore be heavier than the original ore. When this proved not to be the case, the theory had to be altered. It was said that phlogiston was like fire itself, weightless and characterized by levity. Warm air from a fireplace, for example, would rise to the ceiling because being imbued with phlogiston, it was not as affected by gravity. And so on. As theories go, it might seem, to our eyes, a little dubious at best. But for over a hundred years it survived because no could come up with anything better. It also survived because it was a kind of transition idea that had one foot (or hand) in a science of observation and another in the philosophical principles of the Greeks such as the "living fire" of Heraclitus.

Another transition idea that had roots in classical thought was the acid alkali theory. This theory of opposites can be traced back to Empedocles (fifth century) and it asserted that bodily fluids revealed themselves to be either acid or alkaline. It was therefore thought that illness came from an imbalance in the natural balance of these polarities. Bile was seen as alkaline, the heart as acidic, as was saliva. Digestion was seen as a kind of warfare between acids and alkalis that was resolved by neutralization. This particular theory of oppositions did not have as long a career as phlogiston because no one really knew what an acid or alkali were, it was just known that they effervesce when mixed. The idea of opposite forces as a basis for understanding the chemistry of substances did not die out however. It was transformed by later discoveries into the electrochemical theory of bonding. In this theory, as we shall see, compounds were thought to be formed as a union of oppositely charged elements.

Did you notice how the world force crept into the above? It occurs in the innocuous phrase "opposite forces" and as such would normally pass unnoticed. But the concept of "force" did not go unnoticed to the minds of the seventeenth and eighteenth century. This was because a force must be felt to be experienced. This gave it a subjective quality that could not be measured and objectified—a factor that made it suspect in an age when truth was weighed on the balance. Never mind that the concept of force was crucial to the science of mechanics, as Newton determined with his laws of motion. In the eighteenth century—the century of enlightenment when human minds could reckon with God, much less Newton—the empirical philosophers David Hume and Bishop Berkeley speculated on ways to remove "force" from the English language as a "relic of animism."[3]

Animism was the belief that all of nature was embodied with a life force, a belief that was common among many early cultures. By the seventeenth and eighteenth centuries a somewhat tamer version called vitalism claimed organic substances could only be created by the vital force of living organisms. But both views put the concept a force in the non-weighable category of principles and indeterminate causes. For people who wanted to pin down truth and make an exact science really exact such vagaries of the mind were unthinkable. But in spite of all efforts to "disappear" or alter the meaning of force, force seemed here to stay. It was too formulated and ensconced in the fabric of science to be removed without leaving an ugly hole to remind everyone where it had been. (It's a good thing they didn't try to remove Empedocles' love and strife, the two forces that lived at the heart of his natural world. They would have left behind something a lot bigger than a hole!) So force it was and force it remained. People were forced to admit that no matter how many times you recite Newton's formula, $F = ma$ or say the words, force equals mass times acceleration, it doesn't do nearly as well as dropping a brick on the toe to convey the meaning of the word force (with an ouch!). So the empiricists were hard put to force their way out of the force dilemma. And yet there was a solution. Not being able to abolish the subjective aspect of force outright, empirical science did the next best thing. The word became an innocuous abstraction something like the word energy. The hopes were that ultimately it would be swallowed up by a grand Theory of Everything (TOE) after being atomized as a strong force and a weak force—the two forces that hold the nucleus of an atom together. And that is the transition saga of how empirical science got a toehold on the thought of the times. But we are getting at least one step ahead of our story. We need to go back to atoms and look once again at how they too were part of the transition scene.

We mentioned in the last chapter that the two philosophers, Leucippus and Democritus, introduced the concept of atoms. We don't know much about Leucippus except that he inspired Democritus to teach that all existence was determined by Being and Not-Being and that Being was composed of atoms. It might be easier to translate Not-Being into the more current concept of Void. But regardless how you translate it, the philosophy stated that spirit and body alike consisted of finite particles and emptiness, and that the way these particles or atoms mixed with emptiness was what determined the existence we perceive with our senses. Atoms of different shapes would combine and reshuffle the stuff of the universe and thereby cause the different substances and the four elements. Like tiny machines, they were guided and propelled by a world of external forces. Which naturally sounds pretty mechanical and that was the whole point. In fact, such ideas lay at the basis if what became the mechanical view of nature so popular among scientists in the seventeenth century. *But as modern as this might sound, such ideas were not the result of work in the laboratory, neither for the old Greeks nor the new adherents to the mechanical model.* As applied philosophy the principle of atomism was used to explain many phenomena in the beginning days of modern science and because it did so successfully it slowly became an accepted reality. For the Greeks, however, it remained pure philosophy. The concept of atoms, as a philosophical principle, was to free the mind from fear, both of death and of the gods. Basically Democritus said that we are made of atoms and that atoms alone are immortal—even the gods were made of atoms. So there was no need to fear death or the gods. Since all was made of atoms and what atoms did was determined by the laws of nature (physics), we might as well make the best of our atoms and live in equanimity with the natural laws that determine us. In other words, enjoy life to the fullest and let atoms be atoms. Contrary to what some have supposed, this didn't translate into hedonism or to a wanton slavery to the dusty senses; rather, it meant that we learn to enjoy the finer things of life. As we fine tune our atoms to become more like the gods we can lead more perfect lives, lives that would nevertheless end when we died and our atoms, determined by the laws of physics, did what atoms always do. Hence the easy out plan: no wrathful gods to please, no afterlife to prepare for, only atoms to obey as we learn that the best way to go is flow with the laws of nature in a harmonious way. But there was a problem: it was too easy. The philosophy of Democritus left no room for free will. And free will, as any reader of Homer or of Greek history knows, was just as fundamental to Greek character as olives or Spartan gruel. So the next champion of the atomistic view,

Epicurus (341–270 BCE), took Democritus to task and created a philosophy of atoms that allowed for free will. He wrote that "it would be preferable to subscribe to the legends of the gods than to be a slave to the determinism of the physicists."[4] To free himself from the determinism of physical law and still retain the easy out plan, he kept the atoms, the void and the concept of endless atomic motion that determined the nature of substance. To bring the principle of free will to this Epicurus said that atoms would swerve in their motion so that different atoms could bang up against each other and create variations of substance. The different shapes of atoms was not so important; the ability of chance to cause new substance became the deciding factor. And if chance prevailed in the nature of things, we, as human beings, can take our chances, make what choices we will, and determine our own existence. Of course he didn't use the word probability, but if he had he'd be ready for that great leap forward that led to quantum physics. Or to some modern theories about how life happened. But Epicurus wasn't any more interested in creating a new science than his spiritual mentor, Democritus. He only wanted a philosophy that allowed people to live in peace with all atoms, no matter if they belong to you, me or to Joe Diabetes over there in Boeotia. But his ideas caught on. Lucretius was to popularize them in his long (very long) philosophical poem, On the Nature of Things, in the first century CE.[5] The poem was rediscovered and printed in 1473 and heralded by the Italian Renaissance as the next best thing to a new wave of the future.[6]

Now, back to the sixteenth and seventeenth century, to the French philosopher, Rene Descartes (1596–1650), the most influential proponent of atomism of his day. Lucretius had been the topic of some fervent philosophical discussion for a good number of years now but only that. The idea of atoms was in the air but it wasn't till the seventeenth century that they became means of actually explaining phenomena. It was the job of Descartes to help make this transition from philosophy to the laws of physical chemistry. His dictum that nature abhors a vacuum suggested that something had to fill space and atoms were the most likely candidates. His ideas themselves seemed to fill a vacuum: they filled a lack of explanation for how air fills any space in any shape of container. It was seen by the Irish born chemist, Robert Boyle (1627–1691), that a gas expands to fill any space because the particles (atoms) of this gas would simply grow further apart. The gas would become thinner but it would still fill the space. Boyle also explained heat as the agitation of particles or corpuscles and in his voluminous and rambling writings became a spokesman for what was known as the mechanical or corpuscular theory of matter. That is to say that with Boyle the mechanical movement of corpuscles or atoms was a useful way of explaining certain phenomena. (Epicurus would applaud that one!) But it still needs to be pointed out that atoms were still regarded as a philosophical principle that explained a few facts. No one, of course, had ever seen an atom. It would be well into the nineteenth century before the concreteness of atoms was universally accepted among chemists. By then the evidence of things unseen would become insurmountable. But a new step had been taken. The effect of Boyle's writings was to create an understanding between the mechanical philosophers and the chemists of his day and show the philosophers that in order to understand their own philosophy they had to account for experimental findings. The age of empiricism was born.

One might say that it was a twin birth: alongside the atoms and the mechanical model came the mechanical philosophy that made God into the Great Clockmaker who set the universe running and let the cogs and gears do the rest. The philosophy was granted ism status with the name of Deism, and became something of an established creed. Religion or philosophy, Deism offered a mechanical worldview that made nature easy to analyze. Hard facts and certainty replaced the philosophical ambiguities that seemed to characterize the thought of older cultures. But the old can anticipate the new; modern atomism finds its roots in the thought of Democritus and Epicurus. Passed down from Epicurus to Lucretius and then to Descartes, the old clock of atomism was wound and set running for a new age of mechanical model. Boyle set his clock to the new times and set to work applying the mechanical model to a new science. And why not? After all, a science based on tangible facts one could weigh and measure at least had the potential of being certain and solid. Understanding nature became a task of analyzing the cogs and gears of the clock. But what about the human side of the equation? Where did human beings fit in this clockwork universe? Faced with somewhat the same free will problem as Epicurus, Descartes had to set human beings apart from the mechanical laws of physics. To do this he philosophized that only man has a soul; the rest of nature is soulless. His dictum "I think therefore I am" was supposed to show how by thinking we prove the existence of this soul. The ability to think, for Descartes, was what set man apart from the rest of nature. Many would agree with him and this separation between man and nature became known as the Cartesian partition. Boyle and many others accepted this as an adequate view of the universe and for certain physical phenomena like the elasticity and spatial expansion of gases this mechanical model seemed to provide the right answer. But the bridge from philosophy to empiricism proved to be a long one. Boyle never really crossed it. He failed to create a working hypothesis that would enable the chemist to explain what goes on in the fires and alembics of the laboratory. Small wonder: this task was to require efforts of scientists for the next two hundred years.

NOTES TO CHAPTER 2

1. The concept of intelligent design refers to how intricate and complex systems in organisms imply that some degree of intelligence was responsible for the complexity of the system. The term has always suffered in the hands of empirical science because it always seemed to deny the strictly empirical assumptions of modern science by implying that there is some creative intelligence behind the complex relationships that make life happen. But Michael Behe's book, *Darwin's Black Box*, opens the question up from the viewpoint of biochemistry with some surprising insights. We go into these in the last chapter.
2. *Norton History of Chemistry*, pp. 80–82
3. *Goethe's Natural Science*, p. 159
4. Introduction to *On the Nature of Things*, p. xxvii
5. The Presocratics were more concerned with the nature of matter and the Socratic philosophers, Socrates, Plato, Aristotle and Plotinus, more concerned with moral questions of how to live, how to best achieve a perfect society, etc. But it was a unifying trait in Greek philosophy as a whole to express a profound reverence for the mind. This was just as true with the atomists and the Stoics as it was with the Platonists and Aristotelians. Democritus was said

to have starved himself to death when, in old age, he found that his memory was failing him. Epicurus looked upon the refinement of the intellect as the one thing that provided the surest link to the gods. The refinement of the mind provided the greatest pleasures in life. According to Lucretius powers of the mind also seems to hold sway after death. In Book Three of *On the Nature of Things* he states:

> If only they perceived (the cause of their malady) distinctly, they would at once give up everything else and devote themselves first to studying the nature of things; for the issue at stake is their state not merely for one hour, but for eternity—the state in which mortals must pass all the time that remains after death (1071–1075).

This seems to contradict other assertions that body and spirit, being composed of atoms, fall into endless sleep at death. Actually, the point is more that whatever happens after death is not worth thinking of while alive, and most certainly not worth dreading. Death is irrelevant; the study of natural law and the "nature of things" is all that matters. That is what provides peace of mind while alive. We might ask if this might be compared to modern points of view. Or, if it has any relevance to the modern scientific attitude.

6. We might want to ask what Plato would think of the above view expressed by Lucretius. He would probably say that souls that can so forget the spiritual world have drank too deeply of the River of Forgetfulness before being reborn and have thus ascribed to the folly of denial and illusion. (See Book X of the *Republic*) But we are a long way now from Plato and will need to go a lot further . . .

3
A New Science is Born

THE EIGHTEENTH CENTURY has been called the Age of Enlightenment, ostensibly because of the intellectual advances taken by the brightest denizens of this ambitious period but more because it was a time when civilization, for better or for worse, began to look more like our own. In the hands of the encyclopedists, Chambers and Diderot, knowledge became a thing one could quantify and collect. In England, Samuel Johnson edits the first English dictionary. His contemporary, Goethe, pioneered studies in human and plant morphology, developed a theory of light that countered the corpuscular theory of Newton, and wrote his lasting comment on science with his *Faust*. In France, meanwhile, La Metrie writes *L'Homme machine* (The Man Machine), a book that is symptomatic of the times and the extension of the Cartesian view of mechanical nature to include man himself.[1] (Consider how, in our "modern" times, we rather casually compare man to a computer). It was a time of great wit in letters. We have the plays of Molière, the satires of Voltaire, the biting couplets of Pope and the witty dramas of Sheridan and Goldsmith. It was also a time of social unrest and change. The seminal writings of Jean-Jacques Rousseau idealized life in the Nature of the New World and argued just as forcefully in favor of the essential goodness of man as the behavior of the reagents of the period seemed to attest to the contrary. England suffered under the two Georges and lost her American colonies. France endured the despotic legacy of Louis XIV whose long reign lasted till 1715. Subsequent kings did little to address the needs of the people who finally resorted to revolution. The French War of Independence reaped the bitter and bloody harvest of centuries of repression while extolling the ever elusive virtues of *liberté, egalité, and fraternité*. As might be surmised, it was a time when emotions ran high. It was a time when human rights were a hats off topic to some and a heads off topic to others. It was above all a time when the mind and the head that contained it were seen as a highly individual organ capable of changing the world. It was no accident that the century was called the Age of Reason. In pursuit of "reason" a long list of eccentric and remarkable characters would take Western civilization down the rational road of humanism and natural science.

Among these pioneers and innovators, none was more remarkable nor more congenial than Joseph Priestley (1733–1804). Truly a man of his times, Priestley's life was as pivotal as his experiments. Unlike Boyle, who was a friendly, correct and long suffering Anglican dedicating his life to posterity and science, Priestley, a Presbyterian in the days when holy words could be fighting words, was caught in the middle of dissension and even riot. He was honored to be made a citizen of the French Republic for publishing a caustic reply to Edmund Burke's attack on the French War of Independence. In fact it was at a Bastille Day meeting in support of political freedom in general that the dissenting Priestley saw his meetinghouse burned. Though he was to win four thousand pounds in a court settlement against the good people of Birmingham for this deed he continued to suffer persecution during these inflammable times.

Yet Priestley was not a Samuel Adams. He was a very complex man capable of intelligent conversation on many levels and in some half a dozen languages.[2] His inquiring mind was always seeking new frontiers of intellectual as well as political and religious freedom. Destiny seemed to be waiting. It took the form of an encounter with that dapper dandy of Franco-American fame, Benjamin Franklin, to kindle the already smoldering fires of Priestley's scientific calling.[3] With the characteristic enthusiasm of a true *amateur* (in the original sense of the word as a devoted enthusiast), Franklin communicated his zeal for the world of science—a zeal that could well have cost him his life when kite-fishing for lightning. But regardless where he found his ideas, whether in this world or baiting the next, Franklin had been where whole new realms awaited discovery. True enthusiasm is contagious. Priestley caught the sacred flame. With a loan of books from Franklin and research of his own he summarized what was known about electricity and what he had found out on his own in *The History and Present State of Electricity*, in 1767. In the meantime, based on his own work with static electricity, he had been elected a member of the Royal Society.[4]

His first researches in chemistry had a curious beginning at the public brewery next to his house where he experimented with the smelly gas that bubbled to the surface of the large vats. We can only wonder at what his parishioners and neighbors must have thought. A minister lighting wood chops and then watching them go out over the strange gas must have been a bizarre sight for those more accustomed to sermons on demon gin and the sins of Fleetstreet. But he had other interests. He suspected the gas might be the same "fixed air" previously discovered by Joseph Black when heating limestone. He was also attempting to explore the connection between air and spirit. After all, hadn't God breathed life into Adam? Little did he suspect at the time that his search was breathing new life into science. Further experiment showed that the gas made a pleasantly effervescent mixture when mixed with cold well water and allowed to warm to room temperature. The gas wasn't all that soluble but enough dissolved to give a delightfully fizzy taste that reminded Priestley of the authentic Seltzer. Of course he was not aware that he had discovered what we now call carbon dioxide nor was he aware that the sparkling glass of refreshing carbonic acid he offered the Royal Society was the world's first Club Soda. But there it seemed that good taste prevailed. In 1773 the Society awarded him the Copley Medal for his report on fixed air in which he emphasized its possible *medicinal* value.[5] Evidently, with his work on electricity and now, with some respectable fizz to recoup any fall from flavor, he could be passed off as a harmless eccentric if need be. And everybody knows that in the best of times the English love eccentrics. Unless, in the worst of times, one becomes too revolutionary, too outspoken, or just plain too out of line to be a good thing . . .

Meanwhile, in the early days of discovery, the indefatigable Priestley pursued a career as an eccentric priest while his flock wondered at the possible dementia that lay behind his fervent interest in strange stinks and quaint bottles. Among the former, hydrogen chloride and ammonia no doubt exceeded the confines of family discussion. (Priestley was also raising a family at this time). He was able to collect these two gases, both of which are very soluble in water, by putting a familiar invention, the pneumatic trough, to especially good use. As originally designed by Stephen Hales in 1727, six years before Priestley was born, the pneumatic trough provided a way to let gases bubble up into a vat that held an inverted vessel full of water. The gas would displace the water, which went into the vat, and be collected in the jar. But this, of course, only works if the gas is insoluble in water. Borrowing from Cavendish the idea that mercury might do the job better, Priestley set to work and soon collected enough of the two soluble gases to combine them. He was pleasantly surprised by a white cloud that settled to form a white powder of unknown substance.[6] Using the same technique he went on to collect a gas that been discovered a century earlier that had the strange property of turning brown when exposed to air.[7] This gas, known as saltpetre gas,[8] and its ability to brown out and use up the vital part of air—the part that would support plant life—was well known. It was its ability to use up part of the air that had led Robert Boyle and others to question the homogeneity of air. Experiments with these gases would lead Priestley to carry this question to a decisive answer and provide his greatest discovery—a discovery that would enable him to make a far more astounding appeal to the Royal Society than ever his brewery gas and artificial Seltzer.

As with many great events, a setting of fortuitous coincidence aided and abetted the great discovery. Priestley's experiment was hardly new. It was the heating of red calx of mercury to obtain free mercury, a reaction well known since the days of alchemy. What was different was that Priestley, using the pneumatic trough with mercury, collected a substantial quantity of the gas given off. This in itself wasn't so unusual either. The gas had been collected before[9] and even Priestley had performed the old alchemical reaction that produced the gas by heating saltpetre. But he didn't collect enough to test it out. This time, however, enough gas was collected to make the difference. And Priestley happened to have a lit candle nearby and wondering what if . . . he placed the candle in the jar containing his newly collected gas. Unlike his experience with the "fixed air" the flame was far from extinguished. Indeed, it burst into a white flame. A glowing coal from the nearby fire did the same with the added thrill of a few sparks. Here was a wonder of wonders and Priestley was quite at a loss to explain it. Providence works in strange ways. Priestley, writing of the pivotal event some years later in his memoirs, wonders at the coincidence:

> If I had not happened to have a lighted candle before me, I should probably never have had the trial and the whole train of my future experience relating to this kind of air might have been prevented. . . More is owing to what we call chance than to any proper design or preconceived theory.[10]

This chance made history. It enabled Priestley to demonstrate to a world ready to listen that air was composed of not one gas but at least two, one of which would support combustion. Robert Boyle, referring to similar findings, had written that air contains a "vital Quintessence essential to animal life, although most of it serves no such purpose."[11] But like much of Boyle's work, this too had been ahead of his time. A hundred years later Priestley found an audience more ready to accept a challenge to the Aristotelian purity of the four elements.

Air was not an individual element. This much was definitely proven. The nature of the unknown gas Priestley discovered was such that if it itself would not burn it would certainly cause other substances to do so. Priestley thereby called it dephlogisticated air, or air that was so hungry for phlogiston it would take it from whatever substance it could. This dephlogisticated air could sustain a mouse as well as a flame. The story has it that Priestley placed two mice under bell jars containing ordinary air and his newly discovered gas. While waiting for the results he became so engrossed with playing his flute he lost track of time. When he rushed back to see what had transpired he found that the mouse in ordinary air had expired. Deprived of the vital nature of the newly discovered gas, the little creature had given up the ghost. But the other mouse, given a more ample supply of the life sustaining air, was still hail and hearty. Priestley, more attentive this time, ran the experiment again and was able to prevent a second tragedy. Later experiments with lead showed that when the metal is heated in air it too would produce a reddish powder, which, upon reheating, will become lead. By reversing the process he confirmed that metals as well as animals "breath" this vital substance. Like breath, what had been given off had been taken in. The cycle was complete.

The work was far from over, however. From 1770 to 1800 Priestley continued to use his pneumatic trough to study some twenty new "airs." Among his discoveries were (in our terminology) sulfur dioxide, nitrogen and the oxides of nitrogen, carbon monoxide, hydrogen chloride and, of course, oxygen.[12]

Priestley lived not only to see his famous discoveries carry his name abroad, but to finish his days in the country he had defended with such zeal for a righteous cause, America. He came to the new land of hope and ambition and finished out the last years of his life while Cavendish, his crusty contemporary, reaped the rich fruits of a lifetime of discoveries. The principal and most seminal of these discoveries was like a curious twin to Priestley's dephlogisticated air.

Much has been said about Cavendish's notoriously misanthropic ways. Rich where Priestley was poor, aristocrat where Priestley was a commoner, agnostic where Priestley was devout, the two men make a kind of odd couple, opposite in almost every way and yet complementary in their extremes. A fellow founder of pneumatic chemistry, Cavendish's interest in gases set upon the trail of the enigmatic air given off when acids react with metals. Paracelsus had written of this reaction over a hundred years ago; others had done the same. But no one had bothered to collect sufficient amounts of the gas to really study it. Cavendish collected it by placing iron, zinc and tin in dilute oil of vitriol (sulfuric acid) and collected the gas. Then he did the same with hydrochloric acid. Upon testing with a lighted taper he discovered the same blue flame to burn with each sample of the gas. Leaving no ash, causing no smoke, the gas seemed like the very quintessence of fire itself. Then he weighed it and discovered that although lighter than air, it still had weight. There could no longer be any doubt: here was indeed the elusive phlogiston!

He communicated his astounding results to Priestley who discussed them in his *Lunatics Society* where many of the leading savants of the day met to discuss the latest scientific developments. Other discoveries were also discussed there at this gathering of eccentrics and free thinking intelligentsia, discoveries such and the fabulous lightning jar created by the Dutch physicist Pieter van Muisschenbroek, now called the Leyden jar. Pieter demonstrated the effects of his device by lining up nine hundred monks,[13] each connected by an iron wire. The device was charged and great was the jump as the jolt discharged through the obedient hands of unquestioning brothers. Of course the joke has it that here was a leap of faith of a new order. And now, with his discovery of phlogiston, Cavendish instilled a leap faith in the new order of science. Then he created another challenge as he set about to discover what really happened when his newfound phlogiston burned.

After ten years of careful experimenting, using new spark devices to discharge various mixtures of his new gas and the dephlogisticated air of Priestley, he finally came to the conclusion that phlogiston and dephlogisticated air react to form nothing less than water. It is hard today to understand how something so basic could require so much effort and demand such intricate preparation. But Cavendish did not want to leave anything to chance. He was also intrigued at the mixtures of the two gases that would produce the best results. Then there were those pesky errors that inevitably slipped into an experiment due to the crude apparatus of the day. There was the task of weighing the results and comparing this weight to an equal volume of water. This required the best balances of the period. The work was tedious and time consuming. Fortunately, Cavendish had plenty of time and money, being one of the richest men in England, but he was never in any hurry to rush into print with the great news. No, he was a very careful and methodical man who realized that he would have to be absolutely certain before he dared announce that the Aristotelian element of water were indeed a compounding of two substances.

In March 1783, he made his conclusions known to Priestley. The following January he dropped the bomb before the Royal Society. Water? A compounding of two tasteless vapors? Where was the proof?

Cavendish gave a dramatic demonstration of how he came to his conclusions by gathering large quantities of the gas and condensing the vapors given off. Then he showed how two volumes of his phlogiston react with one volume of dephlogisticated air to produce water with no gas left over. Then a curious thing happened. All of a sudden it seemed as if everybody had discovered the same thing. A controversy arose over who discovered it first. News of the discovery comes from Holland and France and other parts of England. To his credit, Cavendish could care less about such squabbles. But it wasn't till some ten years had passed before the issue was decided. The President for the British Association for the Advancement of Science published a lithographed facsimile of Cavendish's original notebook where the dates of each experiment had been duly recorded. This put a rest to the matter. But the controversy brings up an often overlooked but frequently occurring aspect of scientific discovery. Is it just coincidence that brings so many minds to a single focus so that several are discovering the same thing at once? Is it pure chance that Priestley happened to have a candle nearby? Or is there something literally "in the air" as Priestley might have it, something akin to the union of spirit and matter that he was looking for in the first place? The answer to these questions would take us far into the twentieth century but they have become part of the living legacy of science that doesn't fit the mechanical model.[14]

NOTES TO CHAPTER 3

1. Though symptomatic of its time, *L'Homme machine* and its author were condemned by the medical establishment in Paris as well as by religious leaders all over Europe. It was chastised by religious leaders because it took Descartes one step further and said that not only do animals not have souls but neither do human beings. Or, if they did have a soul it was located somewhere in the brain. (As opposed to Van Helmont placing it in the stomach.) But La Mettrie made it clear that he subscribed to the Lucretian view that thoughts, like everything else, were composed of or created by physical causes (atoms) that could be analyzed by physical means. The treatise was condemned by the medical authorities because it renewed La Mettrie's ongoing crusade against academics who refused the new empirical medicine being developed elsewhere in Europe, especially in Holland by Hermann Boerhaave, La Mettrie's teacher and main source of inspiration. Boerhaave was a very influential

teacher of his time who sought to unite (though not without reservations) chemistry and practice while lending strong support to the mechanical model. Underlying the work of both Boerhaave and La Metrie was the quest for certainty that so colored the age. This quest strove to cast out the old school of philosophical vagaries that created such frustration for Paracelsus. In fact, Paracelsus was a major influence on Boerhaave; both men allied themselves with the need for effective cures based on actual practice and both enlisted the aid of chemistry in the effort to find them. La Mettrie polemicized this impulse with satire and vitriolic wit at the expense of the academic school of medicine in Paris and earned a heady place as the most hated man in medical academe. When his *L'Homme machine* appeared he was already half way to the pillory. But vitriolics aside, La Mettrie was doing little more than stating the case for what would become accepted thinking in the nineteenth century. As his intellectual biographer Kathleen Wellman points out, La Mattrie's comparisons go from animals to man rather than vice versa. Man evolved from a primitive inarticulate existence to finally develop rationality. This, after all, is essentially what Darwin will propose a hundred years later. And a hundred years after that and we have Desmond Morris's *Naked Ape*. We might wonder what Paracelsus would think about that! (*La Mettrie, Medicine, Philosophy, and Enlightenment*, pp. 171–186)

2. This is all the more significant since Priestley was reputed to have had a serious stuttering problem. But his brilliant mind made up for it. At fifteen, in order to prepare himself for entering the Dissenting Academy at Devantry, he learned Hebrew, Chaldee, Syriac and some Arabic from a local Dissenting minister. Later he was to learn to be proficient in French, Italian and German and especially Greek. While developing his unorthodox opinions at the Academy, in addition to his heavy study load he had a daily regimen of translating ten folio pages of Greek a day. His skill with languages was to stand him in good stead as he struggled to raise a family as a minister of small congregations of fellow dissenters. He more than once supplemented his meager livelihood with work as a language teacher.

3. The Royal Society was something of a club devoted to the advancement of science. It was acknowledged by the ruling sovereign and membership gave status and recognition to those who had made important discoveries or contributions in the field of science.

4. After leaving the Academy Priestley already showed an interest in science. While teaching at a day school he provided his students with an air pump and a machine for making static electricity.

5. Given what Priestley believed about the relation of spirit and air, the medicinal quality was a serious claim. He made it based on the effect of the solution of fixed air and water. He obviously hadn't tried to breathe it like he would breathe oxygen a few years later!

The History of Chemistry

6. We will discover what this is as a class.

7. This gas, nitric oxide, was discovered by the Cornish physician, John Mayow (1641–79). When exposed to air it would turn into the brown nitrogen dioxide. At the time, however, Mayow ascribed to the nitro-aerial theory that compared lightning and thunder to gunpowder, which got its explosive force from sulfur and nitre (salt peter). Mayow thought that "nitro-aerial spirit" and sulfur were two opposing forces just as acids and bases (a polarity mentioned above in relation to Empedocles that lived on as the two-element acid-alkali theory popular at the time). According to his theory, the fiery quality of sulfur would react with the nitrous particles in the air to produce lightning and thunder to create nitre that would descend with the rain to fertilize crops. Nitre seemed to prove its anti-fire properties (pyrophobic) by lowering the temperature of water. It was also argued that nitre caused snow and hail. Interestingly enough, this view has some merit if we consider the inert and "pyrophobic" properties of nitrogen as a polarity to the fiery and reactive properties of sulfur. (*The Norton History of Chemistry*, pp. 72– 73)

8. The reason why is uncertain but the term owes its origin to the fact that it was found that saltpetre or nitre produced a gas upon being heated. This gas was thought to be "spirit of nitre" or nitric oxide. Later experiments also showed that heating saltpetre would produce oxygen, the two gases combining to make the brown acid forming nitrogen dioxide. This led Boyle to suspect the formation of "spirit of nitre" and cast doubt on the theory that related saltpetre to the "nitre" in the air. (*The Norton History of Chemistry*, p. 74) It might also be noted that saltpetre gas or nitric oxide was also commonly prepared by reacting a metal (usually iron or copper) with dilute nitric acid.

9. To set the record straight, oxygen or what Priestley would call dephlogisticated air, was discovered by Carl Wilhelm Scheele in 1771, three years before Priestley. But Scheele did not publish the news of his discovery till 1777, after Priestley had already made his diphlogisticated air a public phenomenon. But the two men had more in common than diphlogisticated air. They were both lifelong supporters of the phlogiston theory. They were both avid experimenters who cared little for theory or even for quantitative science. They weren't concerned about how much of this made how much of that. Instead both men made copious contributions to science by discovering new elements and compounds. Scheele discovered a variety of acids (tungstic, molybdic, arsenic and hydrocyanic), tasting them as he went. He also made the first phosphoric and hydrofluoric acid. Other contributions include silicon fluoride (SiF_4) and arsenic hidride (AsH_3). Captivated by much of the fiery enthusiasm that we saw in Becher, he was a prolific chemical adventurer whose habit of tasting his discoveries and recording their taste was but another example of the naiveté of the period. Nowadays we let rats do the job but thanks to Scheele someone dared give a taste test to hydrocyanic and arsenic acid so

others might know what they are missing. But he wasn't another Mithridates, who survived, "saturated with poison." He died at 43.

10. From *Crucibles: The Story of Chemistry*, p. 44. This comment is typical for the age. Priestley echoes the disdain many expressed (see note 1 above) toward theory as opposed to experiment.

11. *The World of the Atom*, p. 38.

12. *The Norton History of Chemistry*, p. 104. These include the oxides of sulfur and nitrogen, nitrogen, carbon monoxide, hydrogen chloride and oxygen, ammonia, and silicon fluoride. Other sources give a lower number but Sir Humphrey Davy was to observe that "no single person ever discovered so many new and curious substances." And that is quite a compliment from a man who discovered two alkali metals (sodium and potassium), four alkali earth metals (magnesium, calcium, strontium and barium) as well as phosphine and hydrogen telluride.

13. Like some fish stories, this one may have grown in the telling. This version comes from Bernard Jaffe's enthusiastic telling of the story in his *Crucibles: The Story of Chemistry*, p. 59.

14. There is definitely more to science than meets the eye. One of the most remarkable cases of simultaneous discovery is that of the American and French chemists, Charles Hall and Paul Louis Héroult. Without knowing each other, both scientists discovered the electrolysis process of refining aluminum. But the coincidence doesn't end there. They were born the same year (1863), patented their process the same year (1886) and both turned their discoveries into major industries (ALCOA and Péchiney). They died in 1914, each one aged fifty-one years and one month.

4
LAYING THE CORNER STONE

A LARGE BELL RINGS and a new democracy is born in Independence Hall, Philadelphia. The sound is heard around the world but most clearly in France. Long before France sends her great compliment to the Land of the Free in the form of a lady holding the torch of liberty, the despotic legacy of French kings and despoiling nobility is torn to shreds by the Goddess of Reason, the patron deity of the French revolution. Then fear and years of repression combine to erect the guillotine in her place and the Reign of Terror shocks a complacent Europe into a new awareness of itself. A new struggle is born, the struggle of the people against the autocratic powers that would enslave it. The effects of this struggle would literally turn the world on its head. Those who were below would replace the aristocracy at the top; the spirit of democracy would give power to the people. The echoes of this struggle will be heard around the world. They will resound throughout the nineteenth century in the factories of England and Europe, in the trade unions organized to protect the rights of laborers everywhere, in the politics of rich against poor and vice versa and in the far off streets of colonies and protectorates that were only beginning to shake the shackles that bound their peoples to the purse strings of a European aristocracy. Alongside these efforts to turn the world around came, as we have seen, the efforts of science to distance itself from old concepts of universal truths and seize the day with tangible realities that could be experimentally proven. It was as if the discovery of the new and very tangible profits to be gained from colonies in the new world and in Africa extended its mode of exploration and exploitation to the mentality of science as well. Indeed, we can turn the dictum that "nature abhors a vacuum" on its head and apply it to the modern empirical science of the day. We see that science, with all of its emphasis on the atom, on the parts of matter and on exploring the exploitable aspects of matter, did not happen in a vacuum. It was part of a world change of consciousness that made it part of a much greater whole.

But we are getting ahead of ourselves. Let's go back a bit to the years 1763–1766 when a young French aristocrat, Antoine Lavoisier, was helping a geologist friend of the Lavoisier family map the whole of France's mineral possessions and geological formations. During this tour of the natural resources of France, Lavoisier was particularly interested in the chemistry of gypsum, otherwise known as plaster of Paris. Gypsum has to be heated in order to be suitable for plastering the walls of Parisian houses. This heating drives out water from the rock hard gypsum to give a soft powdery residue that can be mixed with water to become hard and rocklike. Lavoisier found it remarkable that water in the original rock hard gypsum had to be freed from the gypsum before it could be used a plaster with the addition of water. He called the original water in the gypsum "fixed water" and the idea that a substance, in this case water, could be "fixed" or chemically united with another substance stuck in the budding scientists imagination and was to prove the undoing of the phlogiston theory some years later.[1]

In the spring of 1772 Lavoisier read an essay on phlogiston that reported how metals actually gained weight when heated and turned into a calx (oxide in modern terms). Now we will recall that a metal, shiny and bright, is supposed to be rich in phlogiston. When heated and turned into a calx the phlogiston is given off. According to theory therefore a metal should lose the weight of the phlogiston lost when heated and turned to a calx. So the essay Lavoisier read was quite surprising; it suggested the opposite was happening. Then there were other papers saying the same thing.[2] But others recalled the lighter than air phlogiston of Cavendish and suggested that the loss of weight was due to an anti-gravity effect. Still others ascribed to Priestley's view that matter is spirit—a theory or idea that seemed born out by the vitalizing effect of his dephlogisticated air.[3] Now, as more research provided more facts, the face of phlogiston was changing as fast or faster than the facts it was supposed to explain. With such "explanations" in the air it was becoming clear to Lavoisier that the concept of phlogiston had become more of an ideal than a reality and as such had outlived whatever usefulness it once had as a key to understanding the hard facts of chemistry. Lavoisier, remembering his experience with gypsum, began to have other ideas.

Since the truth of the matter must have something to do with weight, the wealthy young aristocrat, who by this time had acquired a fully equipped and very modern laboratory for his day, was prepared to give the question a thorough analysis. Thanks to his position with the Fermes Generals, a kind of Department of Internal Revenue run by aristocrats, he had access to the balance used by the national mint. The use of this very accurate balance in the hands of the very able young scientist was to prove the undoing of phlogiston.

Not right away, however. In 1774 Priestley came to France and in a conversation with Lavoisier revealed his own experiments with mercury and the strange dephlogisticated air that was given off. Prior to this time, Lavoisier had thought that air was liberated when a metal was heated to form a calx much in the same way water was liberated from gypsum. Now he realized it was the other way around: when a calx is heated by itself the strange air of Priestley was given off. But when heated with carbon, as was the case them metal ores were refined, the fixed air of Priestley's brewery was given off. This was also the gas observed when diamonds were burnt. Amidst the confusion, Lavoisier perhaps only saw in Priestley's work another opportunity for more much needed research. In 1772 he said as much when informed of Priestley's experiments with hydrogen chloride and the oxides of nitrogen. But he complained that Priestley's work failed to bring any new insight into the reasons for what he observed.[4] From this we can perhaps surmise that Lavoisier was looking for reasons and explanations to help him understand his own experiments. At any rate he never did credit Priestley's work.[5] What we do know is that after this meeting Lavoisier began to duplicate Priestley's experiments with the red calx of mercury. In doing so, he utilized his accurate balance to weigh the initial calx and the products generated as a result of heating it. He carefully weighed out four ounces (grams were not yet established as the preferred scientific measure) of the calx and proceeded to heat it. He heated it for twelve days, nonstop. At the end of the twelve days of constant heating he carefully weighed the mercury and calx that still remained in the heating flask and the gas that was given off. He discovered that together, the two weights equaled the weight of the calx he had originally placed into the flask when the heating began. Thus the conclusive facts were there to read: a calx was really a metal that was "fixed" with air. He called this air oxygen and the calx became known in the modern terminology as an oxide. Then, to remove all doubt, he reversed the experiment and by heating pure mercury caused it to use up about one fifth of a reservoir of air. Thus he not only demonstrated the process of oxidation, but he also confirmed other estimations that the air contained nineteen to twenty percent oxygen.

By showing how chemistry could be understood in terms of finite and weighable quantities of reactants, Lavoisier laid the cornerstone for what became modern chemistry. From the time of Lavoisier, the door was flung open to establish the composition of different compounds by carefully weighing both the reactants and the product(s) they form when chemically united. A lot of work had yet to be done, however, and the above retelling of the story is very simplified. The truth is complicated by many factors not the least of which was the strange behavior of hydrogen, a gas so unlike oxygen. It also complicated the picture that prior to the work of Lavoisier, no one really had a clear idea what an element was or even what a compound was. This confusion was to last well into the nineteenth century but Lavoisier helped provide the tools and conceptual basis for more accurate analysis. The quest was on.

To spur the search even more, Lavoisier also rewrote the book of chemistry, both figuratively and literally. Figuratively, because he redefined the rules of chemistry and made it a quantitative science that used the laboratory to establish factual weights and volumes as a basis for discovery. Literally, because he simplified the existing nomenclature and established what we know today as the language of chemistry. The renaming of "dephlogisticated air" as oxygen was only a beginning. From there he gave the known elements of his day many of the names we still use. The names attempted to describe the properties of each element. Since Lavoisier thought that the gas given off by the heated mercury calx would form acids, he called the gas oxygen which means acid former.[6] He arrived at this name because of the way nonmetal oxides, such as sulfur dioxide and carbon dioxide, form acids when dissolved in water. Hydrogen, the gas he later proved (along with Cavendish) to unite with oxygen to form water, means "water former." Because Lavoisier threw out the phlogiston terminology used by Cavendish and defined water as the compounding of the two elements, hydrogen and oxygen, Lavoisier is often given credit for discovering the elemental composition of water. Cavendish's work predated Lavoisier's however and, differences of terminology aside, it was finally Cavendish who was given credit for the initial discovery. At least that is the way the English saw it. Of course, if you are French you will feel free to interpret this historic moment as French eyes will see it.

All of this controversy only underscores the basic need of the time, the need for a new terminology. Lavoisier's work continued to address this problem in his *Traité Elementaire de Chimie* (Treatise on the Elements of Chemistry) where he defined an element as any substance that could not be analyzed by chemical means. He gave names to the thirty three elements then known, among which are sulfur, phosphorus, charcoal (carbon), antimony, arsenic, bismuth, cobalt, copper, gold, iron, lead, manganese, mercury, molybdenum, nickel, platinum, silver, tin, tungsten, zinc, lime (calcium), magnesium, aluminum, silicon, oxygen, azote (nitrogen) and hydrogen. Because several of the elements had names that began with the letter "m" alchemical names were sometimes retained. Hence mercury gets the alchemical name hegalium and the symbol, Hg ans sodium the symbol Na for natrum. For traditional reasons, gold and silver retain the old Latinate monikers of aurium and argentium. Lead is called plumbium and copper cuprum. The sign language was still the old one handed down from the alchemists, however. This would not change till Berzelius gave the modern abbreviations. Iron, for example was the circle with an arrow used in astrology to designate the planet mars. Copper was given the astrological sign of Venus. You can well imagine what a chemical reaction must have looked like when written with all of the obscure and arcane terminology which represented different things to different scientists. The Treatise on the Elements of Chemistry came a long way toward creating a simplified vision of chemistry. It was to be used for the next twenty years to rechart the science of the day and prepare for a science that would lend itself ever more effectively to research and to the mineral exploitation of nature's resources. In this way the cornerstone laid by Lavoisier would also be used to pave the way to the Industrial Revolution every bit as much as the invention of the steam engine by James Watt. Applying the mechanical model to matter and the Cartesian partition to nature would make scientific exploration and industrial exploitation co-partners in the effort to turn the crank of the natural machine. Added to that we have a human nature that "gets the thing to go" as it exerts the natural machine to create more power for the benefit of both humankind and human unkind alike.[7]

A biographical note must be added to this story of humankind and human unkind, however. Like Priestley, Lavoisier became embroiled in the political events of his time. Unlike Priestley, however, the outcome was to be more tragic than the burning of a meeting house. Lavoisier had the misfortune of being an aristocrat when to be noble was to be persona non grata during the mobocratic rule of the Reign of Terror. History often proves that when those who have heads abuse them, a state of affaires ensues where those who keep their heads can lose them. Such was the fate that awaited Lavoisier who, along with his father in law, was sent to the guillotine in 1774. A man who had based his science on the use of the analytical balance was overthrown by the unbalanced times in which he lived. He was guillotined while Priestley was en route to America.[8]

NOTES TO CHAPTER 4

1. *The Norton History of Chemistry*, p. 91
2. At this time Lavoisier was pursuing the mystery of burning diamonds. Diamonds, it was discovered, would disappear under intense heat *when the operation was done in an open vessel*. When the diamond was buried in charcoal dust and placed in a sealed container no amount of heat would affect it. Not only that,

but the gas given off when the diamond burned was found to be the same fixed air that Priestley discovered at the brewery. Now consider: diamonds glittered and sparkled and were supposed to be just full of phlogiston. When heated all of this phlogiston disappears and all that is left is some fixed air that is very heavy, much heavier than normal air. This would seem to say that the phlogiston, which is forced out of its diamond (leaving nothing behind), creates a gas that is heavier than air. The fact that this only happened in an open vessel meant that heat alone was not responsible for the change. Therefore, phlogiston reacted with air to make it heavier. And therefore phlogiston must have weight—a weight at least proportional to the original diamond. But now we get these papers that tell of how phlogiston rich metals *gain* weight when heated to form a calx. Phlogiston, which must have weight according to the diamond experiment, is lost by the metal and the metal gains weight? You can understand why Lavoisier was left scratching his head. (See *Lavoisier*, pp. 46–60.)

3. This view of breath as giver of life force recalls the doctrine of vitalism that ascribed to life a vital force. (*Norton History of Chemistry*, pp. 99–111) But it also strikes a common chord with the abstract and essentially lifeless concept that matter is energy. This view is of course derived from Einstein's famous equation $E = mc^2$.

4. *Lavoisier: La Naissance de la Chimie Moderne*, p. 54.

5. Priestley was annoyed at this and protested against Lavoisier's borrowing and objected to gaps in his reasoning. Convinced that it was he who had tipped Lavoisier off on the idea of using mercurius calcinatus (mercury oxide), he implies that Lavoisier is acting out of suspicious motives and expresses his concern that this has led to some poor science:

> After I left Paris, where I procured the mercurius calcinatus above mentioned, and had spoken of the experiments that I had made, and that I intended to make with it, he (Lavoisier) began his experiments kupon the same substance, and presently found what I have called dephlogisticated air, but without investigating the nature of it, and indeed, without being fully apprised of the degree of its purity. And though he says it *seems to be* more fit for respiration than common air, he does not say that he has made any trial to determine how long an animal could live in it. He therefore inferred, as I have said that I myself had once done, that this substance had, during the process of calcinations imbibed atmospherical air, not in part, but in whole. But then he extends his conclusions, and, as it appears to me, without any evidence, to all the metallic calces; saying that, very probably they would all of them yield only common air, if, like mercurius calcinatus, they could be reduced without addition. (*Lavoisier: Chemist, Biologist and Economist*, p. 80)

6. Lavoisier knew the nonmetallic oxides formed acids and this led him to

assume that it was the oxygen that was responsible.

7. A contemporary of Lavoisier, was the industrial inventor, Nicolas Leblanc, who was as much a pioneer in commercial chemistry as Lavoisier was in the pure science. He pioneered a method to make sodium carbonate from sodium chloride, a substance that was used in soap and paper and textile manufacture, dye making and in the tanning of leather. After the French revolution it became impossible to import sodium carbonate and the need for an inexpensive means to manufacture it was in great demand. Leblanc obtained a patent for his process and began to produce sodium carbonate in 1791. The process combined salt (NaCl) from seawater with sulfuric acid to make sodium sulfate and hydrochloric acid (vented off as a gas). The sodium sulfate was heated with calcium carbonate and carbon to produce sodium carbonate, calcium sulfide and carbon dioxide. Unfortunately, either for lack of enough sulfuric acid or financing, Leblanc's plant never attained full production. Competition from others who also claimed to be able to produce sodium carbonate from seawater also undermined his financial prospects. Finally, Leblanc was to end his days by committing suicide in a poorhouse. Thus both he and Lavoisier came to a tragic end. But the story of Leblanc does not end here. As seen from the above chemical process, the process he and others devised for manufacturing sodium carbonate gave off the noxious pollutants, hydrochloric acid and calcium sulfide. A visitor to the south of France in Provence reported in 1820 that the "vapors that were given off from the manufacturing plants blackened and burnt the surrounding region; one would think they were standing at the edge of a volcano." (*Histoire de la Chime*, p. 207) The industrial age had arrived.

8. Priestley left for America in April of 1794 and Lavoisier was beheaded May 8, 1794. But the story has an interesting twist. The Swiss revolutionary leader, Jean Paul Marat had pretensions to become a chemist but several years before the revolution Lavoisier had called him to task for his shoddy science and branded him as a charlatan. This so incensed the soon-to-be pamphleteer of the revolution that when time came to even the score the guillotine was waiting to oblige. The irony of this is that Marat becomes a national hero in the Pantheon of Paris while the name of Lavoisier is not even listed among the great men of the nation. But the French, of course, are not alone in such omissions. We also have American politics to remind us ever more how those who have heads still abuse them . . .

5
LAYING THE FOUNDATION

An Interlude

From the time of Lavoisier chemistry became an exact science. With the intrepid faith of the true believer, scientists embarked on the journey toward the ultimate quest for truth and the conquest of nature. Reason and logic, weights and measure were their tools. Yet, this ultimate faith in himself left man alone, desperately struggling to extract the truth of the universe from the sweat of his own efforts. There was a price to be paid, a price perhaps best described by the nineteenth century scientist and historian Berthelot (not to be confused with Berthollet, mentioned below):

> The world is today without mystery: the rational mind claims to clarify and understand everything; it struggles to give all things a positive and logical explanation and extends its fatal determinism as far as the world of morality. I do not know if the factual deductions of reason will reconcile some day divine predestination and human free will, once so fervently discussed. At any case, the entire material universe is claimed by science, and no one dares any longer to resist the face of this claim. The notion of the miraculous and of the supernatural has vanished like a vain mirage, a bygone bias of another time.[1]

These words were written almost exactly a century after Lavoisier wrote his *Elementary Treatise of Chemistry*. Yet, as despairing as they sound, as we shall see, a strange new mystery will begin to unfold in the very hands of the logic that has done and will do so much to strip away the old beliefs. The bare language of matter itself, as naked as the balance of Lavoisier, will show how nature is only the more eloquent for all her naked beauty.

We can begin to glimpse this mystery if we take a thematic look at what has happened so far. From the earliest time breath has been looked upon as sacred. Curious then that so many men of science who laid the foundations for chemistry began by studying the air we breathe. We know that Priestley was looking for some sort of connection between the Biblical "breath" that raised Adam from the dust and the airs he examined in his laboratory. But such ideas were soon to be seen as old-fashioned. With the fall of phlogiston, the last shred of philosophical science had fallen away. Now the mechanical model and a strategy of facts that could be weighed and counted provided the answers to questions of the day. The strategy of facts would be translated into theory and every analytical tool in the chemist's arsenal of apparatus would test it. Clarity and precision were the needed focus; the analytical approach was here to stay. Under its scrutinizing glare we have seen old concepts and paradigms disappear and be replaced by a world of parts and proportions. We have seen how the air just wasn't what it used to be in the minds of the times. It had been analyzed, weighed and measured. It was almost as though the sacred breath of life itself had been given over to the balance when Lavoisier disproved phlogiston. By the time Berthelot could look back on a century of scientific investigation, the air, fire, the very ground we walk on and the water we drink had become the subject of quantitative analysis. Yet, as T. S. Eliot reminds us in his *Four Quartets*, all things of man and nature coincide with ever recurring themes:

> The stillness as a Chinese jar is still
> Moves perpetually in its stillness.
> Not the stillness of the violin while the note lasts,
> Not that only, but the co-existence,
> Or say that the end precedes the beginning
> And the end and the beginning were always there
> Before the beginning and after the end.[2]

The notes from the violin do indeed grow still as nature turns to atoms. But let us see how this end contains a new song, a new music where, as the poet promises, was always there. And proceed with guarded optimism.

The Birth of Modern Atomic Theory

John Dalton (1766–1844), mentioned above, is generally considered the father of modern atomic theory even though others pioneered and championed the idea many years before him. Boorse and Motz in their history of the atom comment on how

> ... it is a sobering thought that two thousand years of atomic speculation (since the time of Democritus) produced no mind able to formulate atomic theory in questions simple enough for direct experimental answers.... It would be logical to expect that if the greatest minds of two millennia had found no way to question nature in atomic terms, only a demigod might be expected to see where mortals were blind. But no demigod appeared. Instead, the fates sent an unprepossessing country schoolmaster of silent mien and uncouth manners, who, in a chance flash of revelation, caught a glimpse of an open door and passed through.[4]

Now wait a minute. Demigod? Door opener? Where others are blind? What is happening here? Didn't we just read few pages back where Paracelsus, La Metrie, Priestley and Lavoisier were breaking down the doors of knowledge, crying for a more empirical, experiment-driven science? It sure sounded that way. But it seems they were too successful. There was such a welter of new facts that people didn't know what to make of them. The facts were like fish in a mountain stream; they were swimming everywhere. And it wasn't enough to watch them swim; one had to catch them. And the thing to catch a fish is a net. That net is what Dalton provided—he provided a theory to catch the facts that were filling the stream of science. The need was for some unifying and relatively simple pattern of ideas—a paradigm—that would turn a chaos of observations into meaningful order.[4] Gone were the days when holistic ideas like the four elements provided a unified worldview.[5] Now the need was for precision and clarity in the realm of physical truth. And along came Dalton. As the saying goes, if Dalton hadn't existed, someone would have invented him. He became the spokesman who articulated the new paradigm—a paradigm with a long past, as we have seen, fostered by the philosophy of Descartes and by the science of Robert Boyle. About his atoms, Dalton was dead serious—in fact, there seems to be little humor in his rather dry life as a Quaker schoolmaster. Maybe that's why he has such a fixation on studying clouds. Not to say that he didn't have clouds of his own. In fact, it might strike some as amusing that as an inveterate pipe smoker he struck the match to the new smoke of the day. We might even picture him as a pipe smoking surfer who caught the wave of fiery enthusiasm that broke on the shores of hard science.6 A hard science indeed, as Berthelot was to lament, a science with a schoolmaster's rule: Dalton was every inch a man of the empirical cut with a ruling urge to quantify and measure. He owed the success of his theory to the fact that it provided a way to do just that. To this effect, Alexander Findlay gives a more sober assessment of Dalton's contribution:

The most important advance made my Dalton in the development of the hypothesis of the atomic constitution of matter was the introduction of the quantitative factor, and since, according to the atomic theory, the relative weights of the atoms, the so-called atomic weights, are fundamental units of chemical science, the determination of these atomic weights was clearly a matter of importance for the verification and for the application of the theory.[7]

The key phrase is the "quantitative factor." Dalton provided a map for determining the nature of matter not on relationships and qualities but on relative weight. The balance of Lavoisier was to reign supreme as it determined what quantity of this substance reacted with a certain quantity of another substance to produce a predictable quantity of product. Thus the search for exact knowledge increasingly defined chemistry in terms of weight and number. And so we see that weight and number became *the* paradigm for the next step in the coming age of science while chemical and physical properties wait in the wings for a more dramatic entry.[8]

So how did this great step for mankind come to be? As said above, Dalton was a meteorologist. Of course, it is understood that he was an amateur meteorologist. No one was getting paid in those days for such questionable activities as looking at clouds and predicting the weather. Every peasant worth his clod could do that much. What made Dalton unique was the way he went about it. He kept pages and pages of notes about the weather. You didn't start a friendly conversation with John Dalton about the weather, not, that is, if you expected a short reply. Here was a man who could tell you far more about cumulus clouds than you ever wanted to know. A short, color blind Quaker who lived by himself and graded papers, he found clouds to be an exciting diversion to a rather colorless existence. And, of course, for a man who loves clouds, England is a paradise. So what in heaven's name made led him from clouds to atomic theory? In the first place, like all amateurs who make the history books, he asked questions no one else could answer and then started to find some answers on his own. He had read from the current scientific discoveries of Priestley, Cavendish and others that the atmosphere was composed of different gases. He knew from the work of Priestley that plants gave off oxygen and that they took in what we now call carbon dioxide. He also knew what you know, namely that carbon dioxide is very heavy, that it could be poured from a bottle and it would flow downward like water. He knew the approximate composition of the air to be about four fifths azote or nitrogen and one fifth oxygen. He was up to date on everything that was in the air about air. And he asked questions.

The question that haunted him most was why, with all the different gases of different weights in the air, why was the composition of air so uniform? In some ways, the question is similar to the one Robert Boyle asked over a hundred years ago. Boyle, as we remember, asked why air was so elastic and why it filled any empty space. And like Boyle, Dalton chose atoms as an explanation. The simple answer he gave for the question of uniform composition of air was that the atoms of air mixed themselves up by colliding with one another and causing a uniform distribution of the different gases.

This sounds so simple, so logical that we moderns are apt to scratch our heads and ask why no one thought of it before. We have to remember, however, that in Dalton's time the idea that the air was a mixture of gases was still fairly new. In schools across the country the old

Aristotelian model was still taught and that was how most people still thought. It still made sense to think of airiness as a quality that applied to all gases (see footnote 5 below). The new analytical way of thinking was new; the old qualitative way still prevailed. Unless, that is, you happened to be a student in the little classroom of John Dalton.

The world soon filled this little classroom. Though Dalton's theory of the atomic composition of air was untried and was to be far from flawless, it was a workable hypothesis and the feeling ran high if the right answers were forthcoming everything would fit into place. To account for the different densities and weights of the elements Dalton gave the atoms weight. He suggested that hydrogen be given the atomic weight of one, it being the lightest element. The rest of the elements could be seen as multiples by weight of hydrogen. If all matter were formed of atoms like the different gases in air then there was every reason to believe by carefully weighing the reactants and the products they produced one could identify the composition of elements in any given compound.

Dalton's theory of atomic weights was made workable by the law of constant composition and constant proportions, a law that stated that any given compound would always be composed of the proportions of elements. Water, for example would always be H_2O no matter where you found it; it would always be composed of hydrogen and oxygen united in a 2:1 ratio. But as obvious as that might sound, it was anything but obvious in the early part of the nineteenth century when Joseph Louis Proust (1754–1808) struggled with conflicting data and a very incomplete knowledge of the elements. He also struggled to convince his fellow Frenchman, Claude Louis Berthollet (not to be confused with Berthelot, above) was insisted that nature has a freer hand. A very competent chemist, Berthollet was quick to champion the chemistry of Lavoisier. Like Lavoisier he was interested in the perfecting of gunpowder and his discovery of potassium chlorate, $KClO_3$, unfortunately led to the very rapid combustion of a powder mill. But his idea that chemical affinity was proportional to the mass of the reactants convinced him that compounds did not have to combine in fixed ratios. This led him to a courteous conflict with Proust that was only resolved after some mistaken notions of what was a compound and what was not were cleared up. Berthollet, for example, thought that glass was an element. Later, of course, it was found to be an oxide of silicon. Finally, after a long struggle to clarify compounds and establish the elements that composed them, Proust's law of constant proportions was vindicated after eight long years of effort. He exclaimed that

> ... the stones and soil beneath our feet and the ponderable mountains are not mere confused masses of matter; they are pervaded through their innermost constitution by the harmony of numbers.[9]

Here at last was proof that the world was made according to number and proportion. Water would always be two parts hydrogen and one part oxygen; ammonia one part nitrogen and three parts hydrogen, etc. Pythagoras would have smiled and said, "I told you so." Indeed, in the efforts of Proust and others to arrange nature in orderly systems and proportions we hear a distant echo of those ancient Greeks, an echo that resonates with the same fervor for universal truth that lived in the minds of those bygone philosophers. The paradigms, they were a-changin', but the quest for a unifying theme was here to stay.

While Dalton was developing his atmospheric observations into the atomic theory of gases, other scientists were making discoveries that would suggest new ways to test the composition of matter. In Italy, Luigi Galvani (1737–1798) had discovered some years before that frog legs that had been preserved in brine and hung on copper hooks would jerk as though alive when the wind blew them against an iron railing. Galvani concluded that this must be caused by something happening in the muscles of the dead legs. His countryman, Alessandro Volta (1745–1827), upon further examination of the jerking spasms, concluded correctly that the phenomenon was caused by an electric current generated by the action of salt water on the two metals, copper and iron. The legs served as conductors of the current produced by the action of salt water on the two metals. This current caused them to jerk. This discovery led to Volta's great invention, the voltaic pile, or the first battery. He replaced the frog legs with a cloth soaked in salt water or acid. He sandwiched the cloth between two metals, say, copper and iron, and showed how it allowed a "current" of electricity to pass between the two metals just as the frog legs had done. A single sandwich was called a pile. By attaching wires on the two metals he could make this "current" go through a wire. By increasing the number of sandwiches, or the number of piles, he could increase the amount of "current." He devised the idea of current to describe the apparent flow of electricity and the word stuck. Today the whole world speaks of electric current though the analogy is far from a proven fact. The idea, for example, that electrons "flow" refers to how electrical energy fills the space in and around a wire. It is a way or referring to the effects of an electric field. But Volta's metaphor seemed appropriate and it has stuck in our vocabulary like a rock at the bottom of a river.

Soon the voltaic pile was put to work. Volta himself used it to break down water and show how the two gases, hydrogen and oxygen could be generated by the electrolysis of water. In England, Humphry Davy (1778–1829) was to apply a huge voltaic pile to molten salts of potassium, sodium and isolate those metals. He was to do the same with calcium, strontium and barium. Later chlorine and iodine would join his growing list of elements isolated by using the voltaic pile. The success of his experiments with the voltaic pile convinced Davy that compounds were held together by electrical forces.

Davy did not, however, develop this explanation into a theory that went beyond the range of his own experiments. This task was accomplished by the Swedish scientist, Jön Jacob Berzelius (1779–1848). Berzelius, working at the same time as Davy, had also used the pile to isolate the elements cerium, selenium and thorium. Knowledge of Davy's work and the work of John Dalton led him to develop the electrolytic theory of compounds. The theory stated that elements united because different elements had different charges. In the experiments of Davy and Berzelius, the positive and negative electrodes of the pile were placed in molten salts. Metals such as sodium or potassium, because they were attracted to the negative pole, were said to be positive whereas nonmetals like chlorine and iodine, attracted to the positive pole, were said to be negative. Likewise, the electrolysis of water indicated that hydrogen was positive (attracted to the negative pole) and oxygen was negative (attracted to the positive pole). The consistency of these results led to the theory that elements conformed to the nature of electricity and to how opposites attract.[10]

In addition, Berzelius completed the job Lavoisier had begun: he further modernized the nomenclature of chemistry, gave it the language we know today, and further refined the elements by finding their atomic weights. Instead of the old alchemical/astrological symbols still used he adopted the abbreviations using the Roman alphabet while keeping the names used by Lavoisier: mercury was abbreviated as Hg (for the Greek, hydrogyros = liquid silver), copper as Cu (for the Greek, Kupros, the ancient name of Cyprus, the reputed birthplace of Aphrodite), gold as Au (for the Latin, aurum), silver as Ag (for the Sanskrit, argunas = shiny, milky white, brilliant), potassium as K (for the Arabic, kali and Latin, kalium = ashes).[11] Then he attacked the question of correct atomic weights for each of the elements. Like Dalton, he compared each element to hydrogen and by careful and tedious experiment arrived at a table of atomic weights. Berzelius found that the weight of carbon, for example, was twelve times that of hydrogen; the weight of oxygen as sixteen times that of hydrogen and so on. By 1826 Berzelius had identified the comparative weights of chlorine, hydrogen, copper, lead, nitrogen, oxygen, potassium, silver and sulfur at very close to the values accepted today. Thanks to the work of this great pioneer of chemical science chemistry was placed on a firm foundation with a systematic approach to understanding the composition of compounds and the nature of chemical affinity.

Of course he was not alone. The work of Proust and the law of constant proportions described above had done much to make the work of Berzelius possible, as did that of Davy and Dalton. And now another scientist gives a breath of fresh air to the old question of gases. In 1808 Gay Lussac (1778–1850) was experimenting with the electrolysis of water and the ratio of two parts hydrogen and one part oxygen thereby produced. The experiment had already been done by Volta and was nothing new. But the fact that hydrogen and oxygen would be produced in simple ratios intrigued the young scientist. He tried other combinations of gases to see if they would follow the same general rule.

He found that a measured amount of nitrogen gas would always unite with three times that volume of hydrogen, producing ammonia at the simple ratio of 1:3. More experimentation showed the same mathematical simplicity. This led to the discovery of the **law of volumes** which states that *gases combine with each other in simple ratios*. But his discoveries didn't agree with Dalton's theory. He found that one volume of nitrogen combined with one volume of oxygen made *two* volumes of nitric oxide gas instead of the one volume predicted by Dalton. Likewise, hydrogen and nitrogen combined to form two volumes of ammonia instead of one. This was well and good as far as the law of volumes and simple ratios went. But why two volumes of nitric oxide instead of one as Dalton's theory predicted? According to Dalton, one atom of nitrogen should unite with three atoms of hydrogen to form one molecule of ammonia. The same for nitric oxide: one atom of nitrogen and one of oxygen should unite to form one molecule of nitric oxide. This contradiction was not to be resolved for another fifty years. Because the law of volumes was a well demonstrated fact, the contradiction put Dalton's theory in serious question. Atoms were the accepted model but what they did and how they did it were questions no one could answer.

His theory was further compromised by the fact that people weren't yet certain what was an element and what was not. According to Lavoisier's definition an element is supposed to be a substance that cannot be analyzed or broken down by chemical means. But it was by no means certain what fit that definition. Some said chlorine was an element (Davy) and some said not (Berzelius). We recall that the reason Proust had such a hard time proving his law of constant proportions was that his challenger, Bertholet, argued that his reactions with glass as well as with certain alloys did not conform to Proust's law. It would be discovered later that glass was really the compound, silicon dioxide. The sheer number of elements was also seen as a problem. Davy and other chemists found it hard to believe that God would create a universe out of some fifty different building blocks. He had already proved that some of Lavoisier's elements were in fact compounds. He argued with others for a simpler theory of matter. After all hadn't the law of volumes and the law of constant proportions proven that the universe was based on simple ratios?

William Prout (1785–1850), an Edinburgh trained physician, offers a solution. Familiar with Greek philosophy, he recalled the search for primordial substance we know from Parmenides. Aristotle had also concluded that all substances were modifications of primary matter. Prout was struck by the fact that experimental evidence showed that atomic weights were close to whole numbers. He reasoned that this could only be because they were in fact multiples of hydrogen, that hydrogen was the primordial substance Aristotle and others had anticipated in their philosophy. He called this primordial substance, the proto hyle of the universe and made it equivalent to hydrogen. One might say that this is not all that different from Dalton's comparative weights that based the weight of the elements on the weight of hydrogen. But Prout is saying something more. He is saying that the different elements are actually made up of hydrogen. Oxygen, for example, with an atomic weight of sixteen, indicated that sixteen volumes of hydrogen had condensed to form oxygen. Prout's hypothesis, as Berzelius called it, became a launching pad for all sorts of discussion and proved to be a stimulus to research for the next hundred years.

Notes to Chapter 5

1. *Les Origines de L'Alchemie*, p. v. (author's translation)
2. *Collected Poems*, p. 180. We have a similar mood in Keat's "Ode on a Grecian Urn." In either case the jar or urn depicts a design or painting that shows an event happening while yet frozen in time. In that way it "moves perpetually in its stillness." As we turn the jar the scene revolves so that you cannot say this is the end or beginning of it, or find where the end precedes the beginning. "And the end and the beginning were always there, Before the beginning and after the end." We have already seen something of this happening with the themes that keep coming back. We might ask, "Where did they begin?" Or, "Were they always there?" What would Plato say about this?
3. *The World of the Atom*, p. 139.
4. The concept of a paradigm shift was applied by Thomas Kuhn in his *The Structure of Scientific Revolutions* to the consciousness shift that enables scientists to break away from old modes of thinking and develop new ones. Examples would be the Galileo-Copernican change in our view of the solar system or Darwin's theory of evolution. Both of these caused a rethinking of a world

view. Kuhn's book has become something of a consciousness shift in its own right because of the accurate manner in which it revealed the dangers of clinging to old ideas that hinder the advancement of knowledge. As one pinch-hitting wag put it, being a scientist (or anything else) demands that however we play the game, after we hit the ball, we've got to let go of the bat.

5. In his *Warmth Course* (pp. 15–16, 45), Rudolf Steiner points out how the Greeks conceived of the element earth as sharing an identity with all solid substance that by virtue of its solidity was self-centered just as the earth is gravity-centered. Thus the "earthly resides within a solid." Water, on the contrary, is not self- centered, rather it is earth centered. Water is pulled uniformly toward the center of the earth; it does not have a center of its own. This is the unifying characteristic of all liquids; hence their watery quality and their belonging to the element of water or wateriness. Air, on the other hand, has no center at all and flies off in every direction. This is a quality that it shares with all gases; hence all gases are in the element of air. As Steiner goes on to say, it is important that we see how the Greek concept of element pertained to how the qualities of a substance related to the earth as a whole. It was a holistic earth view that did not seek to analyze; rather, it sought to integrate. Ultimately this leads to a view that accepts the interrelations of nature as being intimately related to man. One can see how a worldview that truly accomplishes the feat of relating man to nature and to the cosmos can be very satisfying. It is for this reason that we have attempts to devise a mathematical "Theory of Everything" and reduce the universe into a Legrangian, a formula that can fit a tee shirt. But the reductionist model is doomed to failure in this effort from the start. The math for such a theory is based on an abstract generalization, a paradigm that only exists in someone's head. The rest of the body, the whole human being is left out of the theory that professes to be about "everything." And no matter how many dimensions or other worlds one has in a theory, this will not change. Regardless how you do the numbers, mothers still have to have babies to do the counting. And both parents hopefully will be there to answer questions like "who am I?" If we don't then we are on the verge of letting a reductionist paradigm tell us who we are. And the Greeks might think that a trifle too tragic.

6. Like many men of his time, Dalton took his tobacco seriously. Speaking of an acquaintance he was to have made the remark that the fellow could never be a scientist because he didn't smoke.

7. *A Hundred Years of Chemistry*, p. 14.

8. For example, Dalton thought solubilities of gases were a function of the sizes of particles:
   I am clearly persuaded that the circumstance depends upon the weight and number of the ultimate particles of the several gases: those whose

particles are lightest and single being most absorbable, and others more according as they increase in weight and complexity (quoted in *The Norton History of Chemistry*, p. 143).

9. *Crucibles*, p. 92

10. Note how this is progressing. First we have a theory based on weight and number. Now chemical affinity *as it pertains to inorganic molecules* is being defined in terms of electricity. Though this brief history does not attempt to deal with the complexities of organic chemistry which deserves another "brief history of its own" it is nevertheless important to point out to the students how the conclusions that seemed to pertain to the inorganic world only created more confusion in the organic.

11. *Dictionaire des Corps Purs Simples de la Chimie*.

6
THE SEARCH FOR ORDER

THE SCENE IS SET for the nineteenth century. The world of chemistry is confronted with problems. In the first place, Dalton's atomic theory, though still considered a workable model for reactions between liquids and solids, was not giving the right answers for gases. The electrolytic behavior of acids, bases and salts in solution was yet unexplained. There was no adequate explanation on the atomic level for why solutions of acids, bases and salts conducted electricity. The new science of organic chemistry was a tangle of unproven formulas that made no sense at all; isomers, different compounds with the same molecular formula, were not even suspected. (For example, glucose and fructose, two different substances which both have the same formula of $C_6H_{12}O_6$, are isomers).[1] But by far the most troublesome lack was for a comprehensive ordering of the elements. In spite of Prout's hypothesis about all elements being formed from hydrogen, the many diverse and different elements seemed like the many members of an ever growing family. They might all trace their ancestry back to a common parent, but they are all different, they all clamor for their place in the universe and they are all here to stay. The yearned for simplicity of Proust and Davy seemed lost in the noisy, smelly and unruly goings on in the laboratory. Was nature meant to be this way? Persistent ideals of order and harmony said it wasn't. Equally persistent reality said it was.

Science is full of stories. And we realize by now that science is a very human endeavor. Almost without exception the struggle to bring forth new ideas was a hard one. For years Berzelius was to work in a primitive laboratory before bringing forth the first truly accurate atomic weights. Both Davy and his illustrious student, Michael Faraday, came from humble backgrounds that left them no choice but to struggle for their education. There was also the resistance to new ideas by those who held tenaciously to the old ones. But quietly at home in Turin, Italy, Amadeo Avogadro (1776–1856), didn't quite follow the pattern: he simply waited for the rest of the world to catch up with him.

It was a long wait. History is full of "what ifs" and in Avogadro's case the hiatus between conception and recognition was to witness over fifty years of controversy. Avogadro's conceptions were twofold. Based on the findings of Gay-Lussac, his hypothesis stated that equal volumes of gases (at the same temperature and pressure) contain the same number of atoms or molecules. The logic of his argument is not hard to follow. From Gay-Lussac's work we know that gases combine by volume in simple ratios. From this we can easily see that if one volume of a gas combines with one volume of a different gas atom to atom, then the two gases must have the same number of atoms. Since the gases combine atom to atom, they must produce a compound that has the same number of molecules and the same ratio of atoms per molecule. An example of this would be hydrogen and chlorine as one volume of each combines to form hydrogen chloride gas or HCl. Hence, the ratio of the reactants is preserved in the product. In the case of hydrogen and oxygen, two volumes of hydrogen combine with one volume of oxygen, making a 2:1 ratio and the resulting compound is H_2O or water. In the case of ammonia we combine one volume of nitrogen and three volumes of hydrogen. We therefore expect a compound that reflects this 1:3 ratio and we find it: NH_3 for ammonia. Thus we see that gases combine in whole ratios that agree with the molecular ratios of the compounds they form.

Fine. We now agree that one volume of hydrogen will combine with one volume of chlorine to form HCl, etc. We have an experimentally accurate picture of what the reaction produces *molecularly*. But so far we have said nothing about the volumes each reaction produces. And there is where the problems begin. They begin when we discover (as Gay Lussac did) that one volume of hydrogen and one volume of chlorine produce not one volume of HCl but *two*! Likewise, one volume of nitrogen combines with three volumes of hydrogen to produce not one volume of ammonia but two. And if you are wondering what the big deal is, consider what is supposed to happen in accordance with Dalton's theory. According to Dalton's theory one volume of hydrogen should *combine* with one volume of chlorine to make one volume of the new gas, hydrogen chloride, HCl. Like wise with nitrogen and hydrogen: the gases combine so that one volume of nitrogen attaches itself to three volumes of hydrogen to produce one volume of ammonia. And so on. But that wasn't what was happening. The gases would seem to combine and multiply at the same time. To explain this, Avogadro hypothesized that the gases in general, hydrogen and chlorine included, were diatomic, that is, they normally existed in pairs. Instead of simply O or N for oxygen and nitrogen we would have O_2 and N_2. Thus we have an example of an idea that attempts to fit the facts rather than starting with an idea and attempting to find the facts to fit it. Applying this idea to the facts, we have the following:

$$H_2 + Cl_2 \rightarrow 2HCl$$

and

$$N_2 + 3H_2 \rightarrow 2NH_3$$

This way, in each reaction, we produce *twice as many molecules*. Hence, we have twice the volume. We only have one volume of diatomic H_2 and one volume of diatomic Cl_2, but when these react, they create two hydrogen chloride molecules. And with twice as many molecules we double the volume of the product. The same show goes with ammonia. This shows how the reaction produces two volumes of gas from only one volume of each of the reactants. The facts and the idea are one: reality is served.

Unfortunately, ideas are often stronger than the facts; we would rather see our ideas than the facts that lie before us. The idea of Davy and Berzelius that chemical affinity was electrical in nature insisted that only atoms of opposite charges could combine. Avogadro's hypothesis made no sense from this point of view: the atoms of oxygen or nitrogen would be of like charge and repel each other. Because Berzelius was the dean and first authority of the scientific community in his day, just as much as Avogadro was a virtual unknown, it was the theory of electrical repulsion that held the field. Avogadro had published his hypothesis in 1811. It wasn't until the Karlsruhe conference in 1860 that his countryman, Stanislao Cannizzaro succeeded in presenting a thorough application of Avogadro's hypothesis that differentiated between atoms and molecules—atoms being the single particles of an element and molecules being the diatomic units of gases and the union of atoms in a compound. It was idea whose time had come. Avogadro himself had died four years before.

After 1860, the application of Avogadro's hypothesis, that gases were diatomic and that equal volumes of gases contained the same number of diatomic pairs or molecules opened the way for determining the correct atomic weights of the gases. The latter part, that equal volumes contain the same number of molecules, comes directly from the law of volumes.[2] According to Dalton it works like this: if one liter of hydrogen combines with one liter of chlorine to one liter of hydrogen chloride molecules. Avogadro concluded that this tells us that the liters of hydrogen and chlorine must have contained the same number of atoms. But Avogadro also explains why Gay Lussac discovered that one liter each of hydrogen and chlorine make *two* liters of hydrogen chloride or twice as much as Dalton predicted. He explained this by saying that the hydrogen and chlorine liters contained diatomic molecules of each gas. A diatomic molecule of hydrogen or chlorine contains two atoms, giving us H_2 and Cl_2. The diatomic atoms combine to form twice as much hydrogen chloride, HCl. The same holds true for all the gases—they are all diatomic. Knowing this made it possible to distinguish between that atomic weight of a gas found in a compound and the molecular weight of the diatomic gas. For example, the weight of chlorine (Cl) in the compound HCl is about 35 times heavier than the atomic weight of hydrogen (H). But the weight of Cl_2 is twice that or 70 times heavier than the atomic weight of hydrogen. (Of course the relative weight of Cl_2 is only 35 times heavier than *diatomic* hydrogen.) This was causing a great deal of confusion. But by understanding that the chlorine gas is diatomic it was understood why the relative weights seemed to vary. This provided a means of determining the correct atomic weights of the gases. And with the correct atomic weights for the gases as well as for the other elements, patterns started to emerge.

In 1865, John Alexander Reins Newlands (1837–1898), a consulting chemist of Scottish and Italian parentage made note of the fact that an arrangement of elements by their atomic weights revealed a pattern of familial correspondence between them. The relationship of the elements according to these correspondences was seen to go through repeating groups of seven. This prompted Newlands' "law of octaves" as some analogy between matter and music seemed inevitable.

No.		No.		No.		No.		No.		No.		No.		No.	
H	1	F	8	Cl	15	Co & Ni	22	Br	29	Pd	36	I		Pt & Ir	50
Li	2	Na	9	K	16	Cu	23	Rb	30	Ag	37	Cs	42	Os	51
G	3	Mg	10	Ca	17	Zn	24	Sr	31	Cd	38	Ba & V	44	Hg	52
Bo	4	Al	11	Cr	19	Y	25	Ce & La	33	U	40	Ta	45	Tl	53
C	5	Si	12	Ti	18	In	26	Zr	32	Sn	39	W	46	Pb	54
N	6	P	13	Mn	20	As	27	Di & Mo	34	Sb	41	Nb	47	Bi	55
O	7	S	14	Fe	21	Se	28	Ro & Ru	35	Te	43	Au	48	Th	56
														49	

Newlands' Table of Octaves

Unfortunately for Newlands, there weren't any Pythagoreans on the Royal Society at the time. He was even less fortunate than Avogadro: his law wasn't ignored, it was laughed out of the Society when he tried to present it. This was partly due to a lack of sufficient experimental evidence and partly, as our wag would have it, due to a lack of the right club soda. Without sufficient proof, the coincidence seemed too extraordinary. But where science rushed to judgment, patriotic fervor came to the rescue. After the periodicity of the elements had been confirmed by a German and a Russian, the Very English Society made amends and awarded him the Davy Medal in 1887. Happily, he was still alive to honor his country by receiving the metal.

The German chemist who was to challenge English pride walked out of the Karlsruhe conference with Cannizaro's pamphlet on Avodadro's hypothesis in his hand. Julius Lothar Meyer was deeply impressed by the clarity the "new" hypothesis afforded to the point of exclaiming that the scales drop from his eyes. The German chemist was not long in coming up with his own version of the periodicity of the elements. Meyer not only based his findings on the atomic weights but also on the physical properties of the elements. He plotted the atomic volume (volume of one gram multiplied by the atomic weight) against the atomic weights of the elements. It was due to Avogadro's provision that gas molecules such as chlorine and bromine were diatomic that Meyer was able to find their appropriate places on the chart. The same correspondences Newlands had discovered are here seen very clearly. It is of particular interest to note how the periodicity of the chemical properties can be generated from a consideration of physical properties alone.

I.	II.	III.	IV.	V.	VI.	VII.	VIII.	IX.
	B=11,0	Al=27,3				?In=113,4	Tl=202.7	
	C=11,97	Si=28				Sn=117,8		Pb=206,4
			Ti=48		Zr=89,7			
	N=14,01	P=30,9		As=74,9		Sb=122,1		Bi=207,5
			V=51,2		Nb=93,7		Ta=182,2	
	O=15,96	31,98		Se=78		Te=128?		
			Cr=52,4		Mo=95,6		W=183,5	
--	F=19,1	Cl=35,38		Br=79,75		J=126,5		
			Mn=54,8		Ru=103,5		Os=198,6 ?	
			Fe=55,9		Rh=104,1		Ir=196,7	
			Co=Ni=58,6		Pd=106,2		Pt=196,7	
Li=7,01	Na=22,99	K=39,04		Rb=85,2		Cs=132,7		
			Cu=63,3		Ag=107,66		Au=196,2	
?Be=9,3	Mg=23,9	Ca=39,9		Sr=87,0		Ba=136,8		--
			Zn=64,9		Cd=111,6		Hg=199,8	

Meyer's Periodicity of the Elements

Mendelyev who derived a table based on both the chemical and physical properties of the elements took the next and decisive step. His work was communicated to the Russian Chemical Society in March 1869, in a paper entitled, "The Relation of the Properties to the Atomic Weights of the Elements." The thoroughness of its contents left no doubt as to the years of research that preceded the conclusive findings. He arranged the elements in vertical columns according to atomic weight. As can be seen from his table on the opposite page, the elements

The Search for Order

Reihen	Gruppe I. — R^2O	Gruppe II. — RO	Gruppe III. — R^2O^3	Gruppe IV. RH^4 RO^2	Gruppe V. RH^3 R^2O^5	Gruppe VI. RH^2 RO^3	Gruppe VII. RH R^2O^7	Gruppe VIII. — RO^4
1	H=1							
2	Li=7	Be=9.4	B=11	C=12	N=14	O=16	F=19	
3	Na=23	Mg=24	Al=27.3	Si=28	P=31	S=32	Cl=35.5	
4	K=39	Ca=40	—=44	Ti=48	V=51	Cr=52	Mn=55	Fe=56, Co=59, Ni=59, Cu=63.
5	(Cu=63)	Zn=65	—=68	—=72	As=75	Se=78	Br=80	
6	Rb=85	Sr=87	?Yt=88	Zr=90	Nb=94	Mo=96	—=100	Ru=104, Rh=104, Pd=106, Ag=108.
7	(Ag=108)	Cd=112	In=113	Sn=118	Sb=122	Te=125	J=127	
8	Cs=133	Ba=137	?Di=138	?Ce=140	—	—	—	————
9	(—)	—	—	—	—	—	—	
10	—	—	?Er=178	?La=180	Ta=182	W=184	—	Os=195, Ir=197, Pt=198, Au=199.
11	(Au=199)	Hg=200	Tl=204	Pb=207	Bi=208	—	—	
12	—	—	—	Th=231	—	U=240	—	————

Mendelyev's Periodic Classification of the Elements (1872)[3]

are also arranged according to horizontal groups or periods. Due to its unique properties, hydrogen appears by itself in the upper left hand corner. Immediately beneath it, lithium begins a second period of seven elements that ends with fluorine (F) in Group VII to the far right. Just as in Newlands' table of the elements (which shows the periods vertically), a second period follows the first and extends from sodium (Na for natrium) to chlorine (Cl). The elements of this period demonstrate chemical properties which have a direct correspondence to the chemical properties of the first period. Sodium, for example, is similar to lithium (and hydrogen) in its reactivity. All three elements react with oxygen at a ratio of 2:1 (as in H_2O). The reactivity with oxygen is shown at each group heading by the expression R_2O, RO, where the "R" stands for all of the elements of a particular group. The R2O means that in Group I all the elements will combine at a ratio of 2:1 with oxygen. Sodium oxide, for example, will be Na_2O. In Group II (RO), all the elements combine at a 1:1 ratio with oxygen, giving us MgO and CaO for magnesium and calcium oxides. Furthermore, the elements go from being strongly electro-positive on the left to just as strongly electro-negative on the right. Valence is the measure of how electro-positive or negative an element is. On the far left, in Group I (R_2O), we have what are known as the alkali metals. Because they combine at a 2:1 ratio with oxygen it is said that the alkali metals have a valence of plus one and that oxygen has a valence of negative two. In Group II (RO) we have the alkali earth metals (beryllium, magnesium, calcium, strontium, etc.) which have a valence of plus two. When we get to Group VII (fluorine, chlorine, etc.) we have a predictable valence of plus seven. It is found, however, that the elements of this group combine with hydrogen (plus one) at a ratio of 1:1. This tells us that the elements of Group VII have an effective valence of negative. Hence the RH, which means that one chlorine, will combine with one hydrogen at a ratio of 1:1 to form hydrogen chloride. From Group IV to Group VII we see both a RO rating and an RH rating which shows more clearly the valence of each group and how it corresponds to the group number. What this means is that the chemical properties of the elements in each group are much the same. The fact that each group repeats in cycles of eight establishes a fundamental order or harmony among the elements that had heretofore been lacking.[4]

Not that there weren't irregularities, however. The third period has a run of seven that goes from sodium to chlorine. It is followed by a fourth period that begins with potassium and extends out to a Group VIII comprised of iron (Fe), cobalt (Co) and nickel (Ni). After calcium (Ca) the period comes to an unknown (at. wt. 44), to titanium (Ti) with a valence of four (appropriately in Group IV), to vanadium (V) with a valence of five, to chromium (Cr) with a valence of six and manganese (Mn) with multiple valences that include seven. Iron, cobalt and nickel are placed in Group VIII because they share the same chemical properties. Then the cycle returns to Group I with copper (Cu) and zinc (Zn) behaving very much like potassium and calcium with valences of one and two. The pattern continues. Because each of the elements behaves in accordance with the group it is in, the properties of the two unknown elements that follow (at. wts. 68 and 72) can be determined. Mendelyev called the unknown elements with an atomic weight of 68 and 72 eka-aluminium and eka-silicon and predicted their properties. He did the same with the other gaps in the chart. When these elements were indeed found to agree with the predictions Mendelyev made the chart became an astounding success. It proved itself and proved that the world of the elements and nature in general was ruled by order after all. We can almost hear Proust's exultant cheer for the "harmony of numbers" and perhaps a distant "I told you so" from Pythagoras. From Mendelyev's work we have not only a harmony of numbers but of properties, both physical and chemical. We have the physical property of atomic weight (and atomic number to come later) and corresponding chemical activity working hand in hand with the same cyclical pattern.

All in all the Periodic Table seems to confirm some hidden law that prescribes order in all its manifestations. The electro-chemical properties discovered by Berzelius and Davy point to a fundamental polarity that matches up with the metals and nonmetals of the Periodic Table. We see this in the valences and the reaction patterns of each group. The groups to the far left and right, for example, are the most electrochemical whereas the elements like carbon in Group IV are the least. With a closer look at these middle elements we see that the table reveals not only polarities, but also affinities. The middle position of carbon is of particular importance. As a middle element, carbon shows an affinity for a great range of elements from either side of the Table. Instead of a polar, electro-chemical bond, carbon shares itself with other elements to form what is called a co-valent bond. It is this sharing of energy with other elements that leads to the function of carbon in the building and structuring of life supporting systems. Without carbon, life as we know it would not be possible.

The nature of a bond is reflected not only in the way an element or compound behaves chemically, but also in the structure it forms physically. Crystals are an example. The rigid form of a crystal indicates some degree of polar bonding, especially in inorganic crystals. The crystal's rigidity testified to the kind of bond it contains and to a kind of atomic tension that holds the atoms of the crystal firmly in place. The geometry of a crystal is also determined by the way the atoms fit one another. For each inorganic substance there are therefore a very limited number of ways a crystal can form. Carbon, on the other hand, forms compounds that can vary all the way from the bark of an iron wood tree to the tissue that makes up the supple bodies of animal forms. This is possible because of the bond-friendly nature of carbon that lets it form huge molecules capable of many complex functions. The huge protein molecules and the rings and chains of carbohydrates, enzymes and other organic compounds all contribute to help create a supple body that can adapt to change.

Carbon is responsible for the chameleon like quality of life forms that have to change and appear in so many different forms and shapes. This is true right down to the molecular level. It is because of the many ways carbon can bond that we have the phenomenon of isomerism mentioned above with relation to glucose and fructose. It had been known for some thirty years that for an unknown reason different compounds of the same molecular weight and composition had completely different properties. In 1828 Wöhller found that ammonium cyanate had an identical composition with the urea extracted from a dog's urine (see footnote 1 below). The list kept growing. The periodic table helped explain why this could happen. It was due to carbon's being located in the middle of the chart and due to its ability to combine equally with elements on either side that it was so prolific in making compounds, many of which had the same structure. But the table didn't solve the problem of how to identify the structure of isomers. For example, the different properties of fructose and glucose could be explained by the different structures of the molecules. But a whole new science had to evolve to study what this structure could be. Hence the term structural chemistry, the chemistry of organic molecules and how to identify the ways carbon can bond to create different molecular shapes. The shape of a molecule was just as important as the elements it contained. Then it was discovered that carbon could bond just as easily with chlorine as with hydrogen. This totally contradicted the electro-chemical theory of Berzelius. With carbon a whole new branch of chemistry was necessary. In the twentieth century this new area of explorations would transform the world we live in.

Electrochemistry was alive and well, however, in the solutions of acids, bases and salts. The work of the Swedish scientist Svante Arrhenius (1859–1927) was to create a new theory of electrolysis from the work of the period to explain why these solutions could conduct an electric current while organic solutions of, say, alcohol or sugar, could not. There was clearly something different happening with the inorganic, polarized compounds that was not happening with the organic compounds of carbon. It was understood that there was some disassociation of the electrolyte, of the acid, base or salt in solution *when a current was being applied*. From the time of Volta it was well known that when a current was applied to a dilute solution of water and sulfuric acid oxygen was collected at the positive pole and hydrogen at the negative pole. But what happens when a current is not being applied? The answer usually given was that the salt, acid or base went back to its molecular form. If the electrolyte was salt, NaCl, for example, it would disassociate into sodium and chlorine while the current was applied to the solution but when the current was removed the sodium and chlorine would reunite. It was thought that the current prevented any chlorine from escaping the aqueous solution or the metallic sodium from appearing. Once the current was removed, the molecules came back together and remained together in solution. The re-association of the salts, acids or bases in solution seemed like the only reasonable answer. Otherwise, what would keep the chlorine from escaping as a gas?

This did not suit Arrhenius. He was puzzled over the fact that a more dilute solution of an electrolyte could conduct current as easily as a concentrated solution. Concentrated sulfuric acid, for example, has the same conductivity as a solution of only one part per two hundred. Similar observations were noted for salts and bases. Up to a point, dilution made little or no difference in the conductivity of an electrolyte in solution. This meant that the amount of an electrolyte in solution was not what determined conductivity. Some other principle must be at work than the physical presence of an acid, base or salt.

To help clarify the situation Arrhenius, along with colleagues working on the same problem, called the disassociated parts of electrolyte ions. In a sodium chloride solution, for example, there would be sodium ions and chlorine ions, designated as positive and negative. The positive sodium ions would rest in equilibrium with the negative chlorine ions. The greater the degree of disassociation, Arrhenius reasoned, the less the resistance. From this he concluded that the more dilute the electrolyte the greater the disassociation. From here he made the leap to what seemed obvious. As one story goes, after a sleepless night in May of 1883 his many hours of work with some fifty different solutions finally bore fruit with the realization that ions are not created by current, they are a natural part of any electrolyte solution. It makes a good story, but in truth Arrhenius may have only suddenly realized the meaning of research that had been done by others some years before. However this may be, he was convinced that ions don't need a current to keep them separate. They maintained their own equilibrium as a whole. Not only that, but it was also clear that water itself was included in this whole. Water was also ionized. Why else would oxygen and hydrogen appear when current is passed through a solution of sulfuric acid and water?

Even though others had already come to similar conclusions, Arrhenius' "solution" to the electrolyte question was hard for a lot of scientists to swallow. Independent chlorine ions in solution? Where were they? Why can't we smell them? Taste them? And sodium? If you have free sodium in an aqueous solution, why doesn't it react with the water to form sodium hydroxide? Why doesn't a salt solution, if free sodium ions are really in the solution, test out to be basic? Finally, however, the fact that the theory fit the facts began to win more acceptance. Again we have an idea, much like Avogadro's hypothesis, that lived partly because it fit the facts and partly because a few believing scientists championed it. Slowly Arrhenius' ions won the day simply because they seemed to explain results in the lab. Test after test was run. Yes, a dilute solution conducted a current as well as a concentrated solution. Yes, this must indicate that there are more ions when the electrolyte is more dilute. But why? Why doesn't a more concentrated salt solution conduct better than a dilute one?

Answers to some of these questions had already been given half a century earlier by Michael Faraday. As a result of Faraday's discovery of induction, it was found that any moving charged body creates a field of electric energy around it. And the greater the number of charged bodies or the greater the movement the greater the field. Sodium and chlorine ions, for example, were always charged and always moving in solution. And diluting this solution, it was found, would increase conductivity.

This could only mean that by diluting the electrolyte the actual number of ions or their movement or both was increased. Then came the clincher. Michael Faraday had already shown that the smallest voltage could produce a current in a salt solution. This suggested that ionization occurred without the application of electrical energy. The next step was taken a few years later when Rudolf Clausius asserted that ions were independent of applied current. It remained for Arrhenius to take up the cause and differentiate between "active" and "inactive" parts of an electrolyte. In a more dilute solution, he argued, there were more "active" parts, more ions, than in a concentrated solution. These active parts or ions were what enabled even the smallest current to pass. To say, therefore, that current created ions was to put the shoe on the wrong foot. It was, if anything, the other way around. The moving ions were already creating a field of electric energy that would act as a conduit to any electromotive force applied to a solution. It was the overall effect of an electric field on the whole solution that made electrolysis possible, not the number of ions in solution.

What had become obvious was that it was just as much the field of energy that determined how and electrolyte worked in solution as the material ions themselves. In fact, some would argue even more so. But regardless what one thinks about ions, it is important here that the quality of energy they demonstrate depends on moving bodies responding to and creating a field of force. Here energy is not such an abstract universal; it has a dynamic quality that creates quantifiable effects, whether in electroplating or in electrolysis or in conducting electro-magnetic fields. It is also important that we consider how the phenomenon of electrolytes must be viewed as a relationship of forces that create a continuum of energy. In this direction we will perhaps come to the kind of thought images that inspired Faraday with a conception of how a field of force creates its own space. These were the thought images that helped inspired Maxwell's work.[5] Speaking of them Maxwell writes,

> . . . Faraday, in his mind's eye, saw lines of force traversing all space where the mathematicians saw centers of force attracting at a distance. Faraday saw a medium where they saw nothing but distance . . . Faraday's methods resembled those in which we begin with the whole and

arrive at the parts by analysis, while the ordinary mathematical methods were founded on the principle of beginning with the parts and building up the whole by synthesis.[6]

Perhaps a truer understanding of electrolytes and of ions will represent something of this wholeness as we continue to explore the connections between energy and matter. In fact, it seems that as the old adage will have it, the more we know the more it seems we don't know . . .

However this may be, one thing does seem certain. The more science advanced toward the twentieth century the more it became clear how intertwined energy and matter really are. By this time William Crookes is already exploring what he calls radiant matter. In 1895 Roentgen will discover X-rays. A year later Becquerel will discover that uranium exposes photographic film through a black cloth. And one of his students, a young Polish girl, will turn this strange event into a research project that will lead to the discovery of two new elements and to new secrets at the very heart of matter. She seemed destined for the task. It was she who was mixing chemicals in her cousin's lab in Warsaw, Poland, when Mendelyev noticed her. The Russian scientist was visiting her father, Dr. Sklodowska, who was a professor of mathematics and physics in the high school where young Marja Sklodowska was working in the lab. Mendelyev was impressed by her and promised a good future for her if she pursued her work in chemistry. She was to come to France some ten years later to form one of the most unique partnerships in science with her husband, Pierre Curie. And so we see once more how the science of human events and the human events of science play on and if we listen, we just might catch a few notes of that violin . . .

Notes to Chapter 6

1. Organic compounds abound with isomers. In fact that is one of the reasons organic chemistry is a different world that inorganic. But the two worlds do cross over. Take the example of urea and ammonium cyanate. Both have the empirical formula of N_2H_4OC, or NH_4OCN for ammonium cyanate and H_2NCONH_2 for urea. The two amino groups on the urea molecule identify its relation to the life process though urea is a waste product at the end of the organic line. Nevertheless, when Wöhler synthesized urea by heating ammonium cyanate in 1828 it was quite a sensation in the chemical world. Until that

time the vitalist theory of organic chemistry insisted that organic compounds could not be created from inorganic compounds, that the creation of organic substances required a vital life force. It became apparent with Wöhler's experiment that this was not the case.

2. This can be shown by considering how a volume that contains one atom of nitrogen combines with three equal volumes that contain one hydrogen each to form one molecule of ammonia. Fine: this is in accord with the law of volumes. The law of volumes also tells us that if one increases the number of atoms equally one will increase the volumes equally and the product, ammonia, will increase accordingly, thus preserving the original ratios. Following this reasoning leads us to the conclusion that equal volumes of gases have equal numbers of atoms/molecules. Then we can slip in the diatomic part to explain the two volumes found by Gay Lussac. In this way we see how simple and yet perceptive was Avogadro's explanation.

Later, in the twelfth grade, a study of Avogadro's hypothesis leads to the introduction to the concept of atomic and molecular weight and molar volume. This offers an occasion to introduce Avogadro's number. Originally, Avogadro's number was the number of diatomic molecules in one gram molecular weight of a gas. But it also applies to the number of atoms or molecules in any substance that is composed of discrete atoms or molecules. (By definition an ionic compound does not have single discrete molecules since they are held together by a single field of energy and thus, in the usual way of speaking, represent collections of oppositely charged ions.) So what is Avogadro's number and how was it derived? Avogadro's number was originally determined by electrochemistry. The idea is this: if the ratio between electrons depositing silver on a plate and the exact weight of silver on the plate can be determined, we can determine how many atoms of silver make up that exact weight. That is because in a solution of, say, silver nitrate ($AgNO_3$) it takes exactly one electron to deposit one silver atom. We can write the equation for the plating reaction as follows:

$$Ag^+ + 1e^- = Ag^\circ$$

With the proper apparatus to measure the current (electrons) and amount of silver deposited it was determined that one gram atomic weight of silver (108g) contains 6.02×10^{23} atoms of silver. By definition of gram atomic weight this of course means that one gram atomic weight of any element has 6.02×10^{23} atoms. In the case of gases, however, because they are made of diatomic molecules, we speak of gram molecular weight and gram molecular volumes, or molar volume. And we discover that a gram molecular weight or one molar volume of any gas occupies 22.4 liters and contains 6.02×10^{23} diatomic molecules.

3. The charts for Newlands' table of the elements and Meyer and Mendelyev's periodic tables were taken from *A Hundred Years of Chemistry*, pp. 50–53.

4. Speaking warmly about a lecture given by Eugene Kolisko on the periodic table and why it was arranged in octaves, Steiner goes on to discuss how matter is permeated with music. (See *Warmth Course*, Lecture XIV, p. 108.)

5. James Clerk Maxwell (1831–1879) laid the foundation for the wave theory of electromagnetic radiation. With what became known as Maxwell's equations he calculated the speed of propagation of an electromagnetic wave and discovered that it was the same speed as the speed of light. This led to the conclusion that light was of the same nature as an electromagnetic wave. Some questions still remain regarding this however. For example, why are transparent bodies generally nonconductors of electricity? And why do polarizing crystals polarize light but not an electric current? And last but not least, why is light nonpolar but electricity polar? These are questions to ask your physics teacher!

6. *Spiritual Science, Electricity and Michael Faraday*, p. 25. This holistic view will become the central theme of the last chapter, "Where It's All Going." Faraday's way of seeing with his "mind's eye" is the same as Goethe's archetypal vision. In either case the mind has developed a sense for seeing how harmonic relationships interact to form a collective whole. Stay tuned.

7
AND THE VIOLIN PLAYS ON . . .

IN 1859, ONE YEAR BEFORE the Karlsruhe Conference tackled some of science's most pressing problems and decided on the metric system as the international system of weights and measure, Charles Darwin published his epoch making work, *The Origin of Species by Means of Natural Selection*. Regardless what one thought of natural selection as a mechanical model of evolution, the idea of evolution was "in the air" and became the banner cry of the day. Evolution was another of those ideas that was ready to hit the ever faster tracks of life. Indeed, Darwin had to rush his ideas into print before Alfred Russell Wallace beat him to the punch with a similar theory. Along with the ever-evolving pace of things, progress became synonymous with survival of the fittest, though Darwin himself didn't coin the catchy phrase. In a mood of mind where the sun never sat, Herbert Spencer's philosophy of social Darwinism[1] spread a doctrine of survival of the fittest to justify the expansionist and opportunistic attitude of the English empire. The modern technological age, which arguably began in 1784 with James Watt's invention of the steam engine, was by now going full steam ahead with a full fledged Industrial Revolution fueled by such raw materials as distant colonies and rich beds of English coal. And while the beds of the poor became thinner and the deeds of the rich became poorer, the breath of change was hardly limited to the smog-thickened air of London. All of Europe had caught the evolution fever and increasingly science was seen as the harbinger of a new species of human being, homo technophilius, or technology loving man. Though it was some time before the members of this new evolutionary stream would be seen walking down the street with cell phones growing from the sides of their heads the seeds were being planted for the latest step in the ongoing evolution of the species.

Did I say ongoing? Well, maybe. To date the human genome has not revealed a cell phone gene. Yet, genes or no jeans, one thing is certain: social selection has long ago outstripped natural selection; the mind has left the naked gene—and the naked ape—behind. Social Darwinism has passed the way of the dinosaurs. Or, it has mutated into yet another monster called social indifference. But monsters aside, the attempt to get beyond the mechanical model has created a need for greater vision. At the very least, with the Periodic Table, the cogs seem tuned to a greater rhythm. The elements seem connected to a greater whole. Beyond the cell as beyond the cave, the phones are ringing . . .

. . . And we pick up the receiver and a melodious voice at the other end informs me that there really is music in the way Mendelyev based his Periodic Table primarily on a marriage between physics and chemistry. And the voice goes on to assure me that atomic weights, the backbone of his table, are physical phenomena; they are as non-chemical as comparing the weights of two bags of sand. And before I can get the beat of that one the melodious voice adds that chemistry has always been the science of how substances create new substances when a *chemical* reaction occurs to create new relationships . . . aha! So that's what she's getting at . . . new relationships . . . and my mind drifts to that wonderful dream-vision in the Fifth Day of the Alchemical Wedding of Christian Rosenkreutz where Venus is revealed in all her naked glory on a couch just waiting for somebody to wake her up[2] . . . but before I get to that part the melodious reminds me where I am at the other end of the phone and the next thing I know we're talking about marriage! Wow!

Is this chemistry or what? On one end of the line I get this alchemical message that relationships—the kind that really connect, the kind that, you know, make things happen—those relationships . . . that's what it takes to make chemistry happen on this end of our conversation. Yes, and about this marriage thing . . . and here the melodious voice turns really smooth and adds that, "Yes, you must know darling that we're speaking of how the marriage of the physical and chemical and the successful descriptions it provided for as yet unknown elements really slipped the ring on the periodic finger." But before I could comment on this being a rather strange romance I really did have to admit how the connections fit, how there was a real need for a marriage like this, how it might have been made in heaven . . . And I stood ready to admit that physics and chemistry were meant to be together, though each was different, and I understood why the French say, "vive la difference!" But I knew that the marriage would see some rocky times (the course of true love never being smooth) and was going to need a lot of music to see it through. I knew that because from now on, the relationship between physical properties and chemical properties will become ever closer. But organic chemistry, chemistry that is truly organic, would always be in a class apart. And yet connected in intimate ways . . . And the voice at the other end of the line, the melodious voice that spoke so true, gave a long sigh and before I could ask for her number, said good-bye.

I should have known; such things are beyond number and measure. Such things belong to the imponderables of science that never make it to the chemical formula. But I knew I could never forget the caller with the melodious voice, could never forget the music that makes the connections that really tie the knot . . . And then my thoughts began to float a bit and I began to wonder about that young Polish girl who caught the eye of Mendelyev. Perhaps it was the seriousness with which she went about her routine chores in a chemistry lab, perhaps it was some intuitive hunch of an older and more insightful mind, but if the story of Mendelyev seeing for her a promising life in chemistry is correct, it did indeed prove prophetic. Marja Sklodowska (1867–1834) did choose chemistry for her career. And as a chemist, she would cross new frontiers of physics. She was to become Madame Curie and the discoverer of radium as she and her husband Pierre turned to a labor of love. . .

A touch of romance? Why not? What else but love or insanity would drive a young couple to spend the formative years of their relationship sweating over a pile of pitchblende in search of an elusive element that exposed photographic plates through layers of dark paper? But let's not get ahead of our story . . .

Pitchblende, a name for a certain uranium ore from Austria, had revealed its strange power to the French scientist Antoine Henri Becquerel (1832–1908) quite by accident. He had covered some unexposed photographic plates with black paper in a darkroom and had casually placed a rock of pitchblende on the paper to hold it in place. Much was his surprise he came back later to discover that the plates were exposed directly beneath the rock. The French equivalent to "weird" is the word *bizarre* and bizarre it certainly was. To rule out some freak accident, he tried the same experiment with different kinds of uranium ore with similar results. There could be no doubt about it. Becquerel thought he had found the strange X-rays discovered by Wilhelm Konrad Roentgen in 1896. Only instead of being produced by a vacuum tube, the rays seemed to emanate from a piece of uranium ore to penetrate the dark walls of matter with an unseen light of their own. But there was a catch: pitchblende did a far better job of developing the covered plates than the other uranium ores he tried. There seemed little doubt: some other element besides uranium must be responsible for the extra power of this enigmatic stone.

Here was science waiting to happen. But nothing happens simply because it happens—nothing is that simple. This time science happened where young Marja—she quickly changed he name to the more Franco-friendly Marie—was working in Becquerel's lab in Paris, at the Sorbonne. But the trip to Paris had not been a free ticket. True, she came from a good family with scientific credentials—her father, Dr. Sklodowska, was head of the science department at the secondary school where Marie had first shown an interest in chemistry—but nothing is certain, least of all in Poland which was literally under Russian occupation at the time. It was a brutal period. Occupation is never the way to go; occupation Cossack style is very close to rape pillage and burn. For these reasons young Marja's decision to leave Poland were not entirely scientific. A long history of having Russia as a neighbor had made freedom a sacred word in Poland and it seems that she had joined a group of young revolutionaries. Perhaps it was her father that convinced her that other options could be more fruitful, a point with which she evidently concurred though the option she chose, to study science, was impossible in Poland which denied higher education to women. After six years of work as a governess to save money for going to school outside of Poland she goes to Krakow, which belonged to Austria at the time. She attempted to enroll in the science program at the university but the secretary at admissions laughed at her. Women were not meant to be scientists; it was biologically incorrect. She was told to take cooking classes instead. And that was the lesson that sent her packing for Paris where freedom, the freedom that once rang the bells of revolution, still retained a melodious ring.

France was also a practical choice. Marja had learned French in Poland and she had a sister who practiced medicine in Paris. With a place to live and a life to live for, the young émigré began her studies in physics and mathematics with a will to overcome whatever stood in her way. Revolutionary zeal melted into the iron determination. Finding the distance from her sister's house to the Sorbonne too great, she moved to the cheapest quarter she could find next to the Sorbonne. Living on a budget of 100 francs a month, she spent the little money she had on her education and her room and ate as little as possible, living on high energy chocolate when she could afford it. Coal for her small stove was a luxury she could do without, even in the cold months of winter. Occasionally fainting from hunger, she was nevertheless able to pull through. Call it guts, call it *fortitude intestionale*, call it what you want but she had what she needed. At the end of her strength, she took first place in her master's examination in physics. And went home to Poland to recuperate.

Returning to the Sorbonne the next fall with a modest scholarship, she took up her studies in the laboratory of Becquerel who accepted the Polish girl on the strength of family background and work she had already done in the university. And... like Mendelyev, the older man had taken a liking to the serious young mind that met his gaze. He had just the job for Marie. He set her to solving problems concerning the magnetic properties of metals in solution. Pierre Curie, discoverer of piezo electricity—the electricity caused when certain crystals are subjected to pressure—was also working in the lab.[3] She felt drawn to this serious young man who could work for hours with the crystals he was studying for his doctor's thesis. And they might have heard that same melodious voice, the one that somehow prompted those visions of Venus on her couch . . . But of course we will never know precisely; such things must join the imponderables of the chemistry wedded to physics. All we know is that they were married in 1895.

They were an unusual couple—and prepared for an unusual honeymoon. Becquerel recognized in the young couple just the team he needed to take on the challenges of the hard work necessary to isolate the unknown element that lay hidden in the bizarre uranium ore. At sixty-five, he was too old to do it himself. He was more suitably at the age to offer a guiding hand. Marie was quick to take it and her enthusiasm probably convinced Pierre to choose the more adventurous route to a doctorate. It was to be her route to a doctorate as well if the obstacle of getting enough pitchblende could be overcome. The ore was very expensive. Fortunately, the emperor of Austria heard of the effort and prompted the Joachimsthal mine in Bavaria to part with a ton of pitchblende residue, the part left over after the uranium had been extracted, and send it to the Sorbonne.[4] A quaint gift indeed from a king to a lady but all the better for both sides of the bargain—the Curies didn't want the uranium anyway. They were after the mysterious element(s) that caused the ore to emit an even more powerful radiation. News of the new research had gotten around in other circles as well. Mendelyev, who always kept abreast of the latest goings on in science, got wind of what his young protégé was up to and sent word that according to his Periodic Table there was indeed room for a new element in the alkali earth family below barium. This could be the one the Curries were looking for. The search was on.

The year was 1896. The place was an unheated shed with a leaky roof that had formerly been used as a dissecting room. The task literally amounted to a mountain of hard labor. Or perhaps blind labor might be more to the point. Without knowing the properties of the elements they were looking for or even the kinds of compounds they made (sulfates, oxides, etc.) all methods of extraction had to be tried. This meant leaching out all soluble salts—especially chlorides—and treating it with various chemicals to break down what resisted the water, the acid and the endless toil. Insoluble sulfides were a prime target[5] and finally yielded the prize. To get them solutions had to be treated with hydrogen sulfide, a gas that makes rotten eggs smell like rotten eggs. Then, when one batch was purified and the desired crystals obtained, it was tested for radioactive content. This was done by placing the suspected specimen between metal plates of a condenser. Because a radioactive material electrifies the air and turns it into a conductor of electricity, the amount of radioactivity could be measured by passing a current across the plates and measuring the voltage. After much coaxing, the seemingly endless hours and days and weeks and months of work finally isolated a metallic substance that proved to be a new element which Marie named polonium.

It was a hard call; the atomic weight of the new element (209) was nearly the same as bismuth (208.98). But the chemical properties differed—the chloride of polonium was heavier than that of bismuth and the density of the metal, when finally isolated, was slightly lower. Now work began in earnest—as if it had only been play before—in the attempt to extract yet another element. It was clear that polonium was not sufficiently strong to explain the radioactive power of the ore. And there was the indication given by Mendelyev that a new alkali earth metal was waiting to be found . . .

Finally, a year later, she was able to look upon the chloride of the element she and Pierre had been seeking—an element so powerfully radioactive that one ton of it would keep a 100,000 tons of water boiling for a year. In the meantime, domestic life and the joys of marriage had not been neglected. Two children and two new elements later she and Pierre would share the 1903 Nobel Prize in physics with Becquerel. The two had coaxed the element polonium, named after Marie's beloved homeland, and the elusive and most powerfully radioactive element of all, radium, from the ton of ore. Radium was so powerful it was more than just dangerous to handle. Pierre had his hand so severely burned from demonstrating it while doing experiments in England that he could not hold his knife and fork at dinner. Becquerel suffered a bad burn on his stomach from carrying a very small vial of radium chloride in his vest pocket when going to England to demonstrate the new element to the Royal Society. Like fire from the gods, it demanded that it be handled with respect.

Then, while at the pinnacle of success and happiness, disaster struck. In 1906, only three years after receiving the Nobel Prize, Pierre was struck down by a cab as he was crossing a street. A heavy van from the other direction ran over his head; he died instantly. Stunned by the loss, Marie now had to live for both herself and her husband. She devoted herself to carrying on the work she and Pierre had shared with such devotion. Against all tradition, she was given the professorship position held by Pierre. She picked up where he had left off—with a lecture on the new element polonium. But much work needed to be done in the lab as well. The element radium had yet to be separated from its chloride and tested to see if indeed it was the alkali earth metal Mendelyev had predicted. At long last, in 1910, she passed an electric current through the molten salt and watched the negative mercury electrode. Yes, sure enough, an amalgam was being formed. She collected the mercury alloy and heated it in a silica tube filled with nitrogen under reduced pressure. The mercury boiled off, leaving behind the long sought pure radium. For this work she won the 1911 Nobel Prize in chemistry.

The work of Marie Curie was to be as seminal in its own way as the work of the pneumatic chemists, Priestley, Cavendish and Lavoisier so many years before. Just as these pioneers had proven that the air we breathe and the water we drink was not what they were once thought to have been, so the Curies proved that matter itself was not the solid stable stuff it was once thought to have been. There was some force, some secret fire, within matter itself that was revealed in elements like uranium, polonium and above all in radium. Did all matter contain this fire? If so, then why did only a few elements reveal it while others remained silent and cold? And what of the X-rays discovered by Roentgen? Where did they fit in with this fire?

Some years before Roentgen's history making discovery, Sir William Crookes had coined the term "radiant matter" to describe the strange radiations that were given off by vacuum tubes when an electric discharge passed through them. Known as cathode ray tubes, they gave off a ghostly light that would later lead Roentgen to the discovery of X-rays and the pictures of the human skeletons that became a worldwide sensation. Though a famous scientist and discoverer of the element thallium, Crookes went the other direction; instead of bones he looked for a more ethereal revelation. Caught up in the spiritualism that was sweeping Victorian England, he sought in the strange emanations a doorway to the spirit and a possible way to contact his deceased brother. Today we might smile at such a naïve notion. As Berthelot had foreseen, the modern, rational mind has no place for such intimations of the supernatural. In our rush to have all the answers we forget that wonder—even the wonder caused by so-called ignorance—is the herald of discovery. For Crookes, however, such considerations simply meant keeping the doors open. While keeping alive a belief in psychic phenomena he went on to analyze the cathode ray emanations. He was convinced that their "fourth state of matter" was particulate in nature. He very pointedly referred to the "projected molecules" in the tubes. He concluded one lecture with the prophecy that his tubes would "reveal to physical science a new world."[6]

The one to discover this "new world" was J. J. Thomson. It is hard to understand the magnitude of Thomson's work unless we realize that prior to his experiments with the nature of cathode emissions the atom was still considered to be the "hard, impenetrable, movable particles" so described by Isaac Newton over two hundred years ago. Thomson, however, was mindful of the work of Michael Faraday. In 1834 this discoverer of the magnetic field had written a paper on the equivalents of electrochemical decompositions. In it he noted that the amount of electricity required to break down a given quantity of water by electrolysis equaled the amount given off when this same quantity of water was decomposed by chemical means (say, by sulfuric acid). Whatever this energy was, it acted the same in both electrical and chemical reactions that left the reactants essentially the same. This energy could come and go, but hydrogen remained hydrogen and oxygen remained oxygen. And yet the energy belonged to each. In the first case (electrolysis) it separated the two elements; in the second (as with an acid) it was given off when they were separated by chemical means. In either case the energy could be measured (in joules, the unit of energy). Faraday concluded that this energy could only come from the "grains" of water themselves. This could only be true, he reasoned, if there was some indivisible and minimal unit of energy/charge shared by these molecules that was taken and given equally. There was even more to the atom than didn't meet the eye.

Faraday went on to estimate the ratio of hydrogen's mass to this charge, e, and discovered the $M/e = 10^{-4}$ or $M/e = 1/10,000$. "What an enormous amount of energy," he wrote, "is required for the decomposition of a single grain of water!" But what was this charge? Did it have mass? Was it, like Crookes thought, some kind of particle? Or was it ethereal—a mere wave of energy somehow connected to an atom? Faraday did not have the tools at his disposal to answer such questions.

Thomson felt there was a connection between the mysterious glow in the cathode ray tube and the Faraday's grains of charge. Like Crookes, he suspected that the tube contained "projected molecules" of energy. Only instead of psychic phenomena, Thomson saw in these strange emanations a more universal truth. As Thomson pondered over the nature of whatever caused the cathode rays or the enigmatic fourth state of matter, the answer seemed to point again and again to some aspect of matter that was common to all substance. Like Faraday, he saw the same energy whether it belonged to a hydrogen atom or an oxygen atom. From Berzelius and others he also had the whole tradition of electrochemistry to suggest that atoms in general were charged. And then there was Arrhenius and his ions ... With all of this in mind, Thomson took the logical course and decided that whatever the cathode discharge was, it must be part of all atoms, a property of all matter. And to find out what it was, he knew he had to allow for both energy and mass considerations and treat the beam as though it were composed of particles, or corpuscles of both energy and mass. The next step took him to the lab where the apparatus was simple but ingenious.

In order to create an experiment that would allow him to determine the nature of the electric emission that glowed across the cathode tube, Thomson designed the tube so he could investigate the effects of both electric and magnetic fields on the discharge. The cathode (labeled as

C on diagram) would send a concentrated discharge over the plates D and E, which could be charged to deflect the beam. This deflection would be carefully measured as it caused a spot of light to move on the phosphorescent area at the bulb end of the tube. The magnets, which would be placed on either side so that the direction of field would be perpendicular to that of the electric field, are not shown.

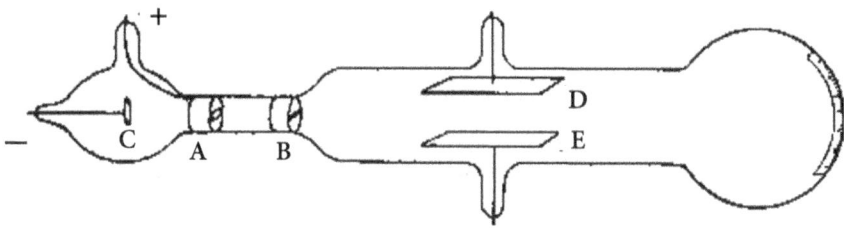

Thomson's cathode tube: The rays from the cathode C pass through a slit in the anode A, which is a metal plug fitting tightly into the tube and connected with a ground. After passing through a second slit in another grounded metal plug B, they travel between two parallel aluminum plates, D and E. They then fall at the end of the tube where they create a well defined patch of phosphorescence.[7]

The electric field between the two plates would indicate both the magnitude of the beam's charge and whether it were negative or positive. If plate E, for example, were positive and plate D negative, this would deflect the electric beam down if it were negative up if positive. The amount of deflection would depend on the amount of charge on an electric particle. As the voltage across the two plates would be known, and the amount of deflection could be measured, this would lead to a way to measure the charge on the electric particle relative to its mass. Don't forget. If the particle has mass it also had momentum. This too affects how much it bends. The result of these two factors gives a ratio of mass to charge that Thomson calculated to be 10^{-7} or a thousand times less than the amount Faraday had found for the ratio of the charge to a hydrogen atom. This meant that either the mass of an electron (as the particles were later to called) was very small or the charge very large or something in between. It would be later confirmed that the mass of an electron is 1/1700 the mass of a hydrogen atom.

And it had a negative charge.⁸

When he had successfully determined the value for m/e with the implications that the likely mass of an electron was very small, Thomson was quick to put forth the idea that the "corpuscles" of cathode rays were a universal constituent of all matter. He writes that

> The explanation which seems to me to account in the most simple and straightforward manner for the facts is founded on a view of the constitution of the chemical elements which has been favourably entertained by many chemists: this view is that the atoms of the different chemical elements are different aggregations of atoms of the same kind. In the form in which this hypothesis was enunciated by Prout, the atoms of the different elements were hydrogen atoms; in his precise form the hypothesis is not tenable, but if we substitute for hydrogen some unknown primordial substance x, there is nothing known which is inconsistent with this hypothesis . . . ⁹

Prout was not alone: Crookes had also used the word protyle to designate this universal substance. But neither Prout nor Crookes were able to isolate a likely candidate as an experimental fact. Thomson's work made the concept of a protyle scientifically tenable. Parmenides, here we come.

It's not quite back to the future yet; but we are on our way. The next major step toward discovering the constitution of matter was to be taken by a student of Thomson, Ernest Rutherford (1871–1937). Rutherford, a native of New Zealand, had come to London in 1895 just as the Cavendish Laboratory at Cambridge where Thomson made his momentous discovery opened its doors to accept research students from other universities. This allowed Thomson to welcome the aspiring young man to be a member of the university. Rutherford joined Thomson in studying the ionization of the air produced by X-rays. When X-rays would pass through the air, they would cause the air to become positively charged as though they had torn something away from the atoms of oxygen and nitrogen. Thomson thought rightly that the powerful X-rays were stripping away electrons, leaving behind the positively charged whatever that remained. This and subsequent work would result in his "plum pudding" model of the atom that stated that electrons were stuck in an atom in a way that resembled cloves stuck into a plum pudding. The X-rays would simply knock the electrons, leaving the positive "plum pudding" behind. Some found this rather amusing and Thomson himself probably saw the humor in atoms having an English flavor. After all, England did lay claim to a considerable portion of the world in those days. Why not claim atoms as well?

However that may be, Thomson was well known for his playful spirit. As for Rutherford, playful or not, the nature of the pudding remained a question to be pursued.

But first a little history. Because Rutherford's effort to get to the "plum" of the matter parallels the work of Thomson, we need to see where the plum starts and where electrons end. Experimental work pertinent to Rutherford's research on the atom began when Norman Lockyer discovered mysterious gas in the sun's corona in 1868. He made this discovery with a spectrometer when he found a spectrum line unknown to any earthly element. With this he concluded that he had indeed discovered a new element which he named, appropriately enough, helium, after Helios, the Greek god of the sun. In 1895 the Scottish chemist William Ramsey (1852–1916), famous for discovering the noble gases argon, krypton, xenon and neon, takes up the study of an unknown gas given off by the uranium rich ore, cleavite. He was able to show that this gas has the same spectrum lines as the gas Lockyer discovered.[10] (But it is important to remember that *this is a gas given off by a radioactive element.*) In 1910 he would also discover the radioactive noble gas, radon, given off by radium. The plot thickens.

In 1899 Rutherford, working with uranium, discovers two kinds of radiation given off, one with high penetrating power that goes through several thin sheets of aluminum and one with low that is easily blocked. He reasons that both emanations must have mass, otherwise they wouldn't be blocked at all. The one with the least penetrating power, having the greatest mass, he calls the alpha particle (α particle) and the one with the greatest penetrating power he calls the beta particle (β particle). To really prove they have mass, he tests them in a magnetic field and yes indeed, they are deflected. One goes the positive way (α particle) and one goes the negative way (β particle). Putting two and two together at this point isn't hard. Ramsay finds helium around uranium;

Rutherford finds the heavy α particle around uranium . . . could there be a connection here? Rutherford suspects there is and goes on record as stating that the α particles are indeed related to the helium atoms Ramsay found. But of course, suspicions don't amount to much without proof. Finally, in 1908, Rutherford gets the chance to perform the experiment that connects the α particle to helium. He puts some radium emanation (radon) in a thin walled tube. The walls are so thin they will allow α particles to go through them but not diatomic helium. The thin walled tube is then placed inside a tube with thicker walls to collect the α particles given off. These he tests with an electric discharge and sure enough, he gets the spectrum lines of helium. The next step was to determine the mass of the α particle. Once the charge was determined to be twice that of a hydrogen ion, and the ratio of charge to mass was found to be 1:2, it was seen that the mass had to be four times that of hydrogen. That put the plum in the pudding. The α particle was found to have the same e/m as an electron.[11]

All of this neatly steps over the years between 1898 and 1907 when Rutherford was professor of physics at McGill University in Montreal. The other half of the puzzle was taken up there, the half that treats what happens to the atom when it loses an α or β particle, or both. With the help of the chemist Frederick Soddy, Rutherford conducted a series of experiments that examined the radioactive properties of various elements, including the radium of Marie and Pierre Curie. From these studies he was to discover that all radioactive elements have what he termed a "half-life" or a definite period of time in which the element loses one half of its radioactivity. The half-life could be a matter of minutes, days or years. For uranium the half-life is 4.5×10^9 years and for plutonium, 24,400 years. Einsteinium, an artificially created element, has a half-life of 276 days and Lawrencium, another artificial element, a half-life of 8 seconds. Radium's half-life is a short 1,620 years, a fact that would account for its being so rare. Half-lives, however, remarkable as they may seem, were not the most astounding part of Rutherford's discoveries. When a substance decayed, it turned into a completely different substance with different atomic weight and different chemical properties. When naturally occurring, radioactive decay begins with uranium. Uranium (U^{238}) loses an α particle (atomic wt. = 4) to make an isotope of thorium (Th^{234}). Then other changes not understood at the time lead us to actinium and radium. When radium (atomic wt. = 226) gives off an α particle it turns into the radioactive gas radon (atomic wt. = 222). Now we find that actinium, polonium and bismuth all have radioactive isotopes with an atomic weight of 210. This indicates the loss of four α particles.[12] Then the last step in the decay process brings us to the stable lead isotope, Pb^{206}. More research will fill out picture and explain intermediary steps (such as the decay of thorium to produce protactinium) but for now it was clear that the loss of α particles largely determined radioactive decay. Here was something close to the alchemist's dream: a true transmutation of substance. Had Rutherford found the Philosopher's Stone?

Well, that depends on how you see it. Some might say the α particle was at the very least a close second. Alpha particles not only became known as Rutherford's pet but they became a runner up for the primordial substance award. They were looking more and more like the building block of matter. Very curious how many isotopes of the elements seemed to vary by units of four. . . Meanwhile, with the jury still out on that one, Rutherford trained his pet to perform its most sensational feat yet. It had been observed that when a narrow beam of α particles passed through a thin sheet of substance the beam becomes diffused. The thicker the sheet the more the diffusion. This scattering was attributed to the atoms of the target material as they deflected the alpha beam. One thing led to another as the alpha beam continued to bombard different targets. Then history happened. The alpha beam actually bounced back from the target—an actual recoil from a head-on or nearly head-on collision! Rutherford's comment was that this was like firing a 15-inch projectile at a piece of tissue paper and having it come back at you. Further experiment confirmed, by increasing the thickness of the foil, that an increase in backwards scattering would result. This proved beyond all doubt that the particles that had penetrated the target were being turned back. For Rutherford, who knew the tremendous energy of the alpha particle, a very massive projectile by atomic standards moving at 10,000 miles a second, this was a phenomenal event in the highest order.

Two conclusions were forthcoming. One, that the atom was largely empty space that would allow the penetration of the alpha beam. Two, that the atom itself must have a very massive core that was capable of reversing the energetic bombardment of the alpha particle. Both of these conclusions were formulated in Rutherford's model of the atom. Though at first he was not certain if the nucleus was positive or negative (in fact, he originally opted for the latter), he was certain that the atom must be the seat of an intense electric field in order to produce such a large deflection at a single encounter. Later, the positive charge of the nucleus was established and Rutherford's planetary model of the atom emerges with a positively charged nucleus and a swirl of electrons outside it. The positive charged particle that helped make the nucleus so massive was called the proton. Because it was found that each hydrogen atom had one proton in its nucleus, the weight of the proton was established as one. You might say, wait a minute—isn't that the same weight as a hydrogen atom? Yes, it is. But a hydrogen atom turns out to have a proton for a nucleus with one electron orbiting around it. Because the electron weighs so little (1/1700 the mass of a proton) the proton is verry verrrry close to having the the same relative weight of the hydrogen atom. Together, the proton and electron, for all practical (and atomic) purposes, could be said to have the weight of one. But this picture was not complete. Rutherford was never able to explain why the atomic nuclei of helium and other elements weighed more than the number of positive particles or protons they contained. A piece of the puzzle was still missing.

The missing piece was to be found, fittingly enough, by one of Rutherford's students, James Chadwick. By 1920, Rutherford had become so certain that some neutral particle had to exist that, typical for him, he began to speak of them as though they already existed. He coined the word neutron and started a search for their existence. It wasn't until 1932, however, that Chadwick, when investigating the bombardment of beryllium with alpha particles, noticed strange rays given off from the target after the bombardment. To test these rays he directed them at a nitrogen target and measured the amount of recoil of the target nuclei. To his surprise, the calculations confirmed the prescience of his former teacher and the neutron was born. It was found that the neutron was a proton (+1) and electron (-1) combined so that the positive and negative charges neutralized each other. Because the electron is so light (1/1700 of a hydrogen atom), the neutron was also given the atomic mass of one.[13]

The neutron, composed of an electron and a proton, is a kind of collapsed form of the proton/electron polarity we also find in the hydrogen atom. Being the simplest element, the nucleus of hydrogen is simply a proton with a 1+ charge. To balance off this positive charge, the hydrogen nucleus has one electron(1^-) orbiting around it. Helium, however, contains both protons and neutrons in its nucleus. It has two protons and two neutrons, giving the nucleus a net 2^+ charge, balanced off by *two* orbiting electrons. Soon it was found that the atomic nuclei of all the elements were made of protons and neutrons with enough electrons in the outer orbits to balance off the positive protons in the nuclei. Together then, *the proton and electron, whether they appear as a hydrogen atom or as a neutron, can be said to be the primary constituents of all matter.* Above we asked if helium might be the fundamental building block of matter. But because helium turns out to have a nucleus made of the protons and neutrons, we can see that the proton/electron pair win as candidates for Prout's protyle (proto hyle) or fundamental particle. True, later research will reveal even more fundamental parts of the atom, but this is pure physics.

As far as chemistry is concerned, hydrogen keeps its primordial rank. According to present theory, stellar fusion causes hydrogen and other elements to combine in a process that forms the other elements. Bringing this cosmic fire down to earth, hydrogen also recalls the phlogiston of Cavendish and by dint of our new findings, the universal fire of Heraclitus. Remember that Chinese jar T. S. Eliot was so fond of? And how it's hard to find where all this ends and begins?

While we are meditating on how the beginning is always here, or there, or anywhere and how the "end precedes the beginning" we might also consider Einstein's equation $E = mc^2$ which translates to say that energy equals mass times the speed of light (in centimeters per second) squared. The point is not just that a little mass amounts to a lot of energy. With this in mind, Werner Heisenberg, noted among other things, for his contributions to quantum physics, also notes in his *Physics and Philosophy* that "energy" is to modern physics what fire was to Heraclitus. He allows that for Heraclitus fire was not an abstraction; like energy that expresses itself in particle and wave or in an endless array of organic systems, it too was a universal that expressed itself in all conditions of existence in different ways. It was what caused the dynamics of "strife" between all bodies and emotions. It was also what caused the life of the mind to respond to an outward flame with a force of its own. It had more forms than Proteus as it lived in a world that was at once one and many. The fire of Heraclitus was both Being and Becoming; it was imperishable change that renovates the world. Acknowledging this, Heisenberg states that "we may remark. . . that modern physics is in some way extremely near to the doctrines of Heraclitus. If we replace the world "fire" by the word "energy" we can almost repeat his statements word for word from our modern point of view."[15]

He also notes the step toward materialism take by Empedocles whose polarities became fixed principles. And then the fire became the deterministic atomism of Leucippus and Democritus. No matter how we turn the jar it seems that the world of energy comes in endless patterns. But one thing we can note as a change, modern science has harnessed energy to the plow of everyday life in a way the Greeks never dreamed. One effect of this is of course to make us complacent as we lazily flip the switch to "get the thing to go." And if it blows up, hopefully it will be on T.V. But another is to think of all the work that got the switch there in the first place. We seldom ponder the pioneering efforts and the years of research that went into creating the electric field that toasts our toast or cooks our scrambled eggs in the morning before we head out of our cave and into the world . . . But maybe we should. Maybe we should even throw in a little philosophy—even a little primordial fire—to spice up our sometimes scrambled life. Maybe we'll get so we like the sound of violins in the morning . . . or something similar to get us in tune with the fire that enlightens us before *we* become toast . . .

NOTES TO CHAPTER 7

1. Herbert Spencer (1820–1903) was the father, along with Walter Bagehot, of

social Darwinism and the person who coined the phrase "survival of the fittest." He was a proponent of evolution before Darwin and argued that differences of rich and poor were a natural result of superior and inherent moral traits, the rich getting richer because they are better. Because attempts to reform society would interfere with this "natural" process of selection, they should be avoided. Free market, unrestricted competition and a laissez faire status quo are therefore in accord with natural selection. Similar states of "mind" have appeared throughout history to justify imperialist and racist policies in one way or another. But social Darwinism is unique in the way it claims biological justification for its doctrine of power and exploitation.

2. The *Chemical Wedding of Christian Rosenkreutz* tells of how the protagonist, Christian Rosenkreutz, is initiated into the mysteries of esoteric alchemy. The allegorical story is divided up into seven days, each day representing a certain stage of initiation. On the fifth day Christian enters an underground chamber after passing through an underground corridor lined with precious stones. Led by his guide, who can be seen as representing his ego, Christian discovers a chamber where Lady Venus is lying naked on a couch. After this revelation, he reads that when she awakes she will be the mother of kings. Aside from a certain romantic flavor this story lends to the above, it is also significant because the work with radioactivity, led by the Curies and followed by Rutherford, will indeed strip away the veil from nature to reveal her in the "nude." But placed in the context of the initiation Christian experiences, this beholding of the nakedness of nature takes on heightened meaning. In what Venus represents he discovers a universal play of forces and relationships that nurtured one another in a way that keep the whole of nature alive. The role of human beings is to wake Venus with the individual consciousness that brings love to the nature and nurture relationship. Enlightened to the higher meaning of love he perceived the inner relationships that create harmony between man and nature, that create a music of the mind that understands how to connect with the flow of things. This wisdom enables Christian to undergo certain trials on the last day that enable him to see the true science that transforms the human spirit and prepares us for a new consciousness and a new relation to nature. And there is where we are today—at the threshold of a new relation to nature. After the discoveries of the Curies evolved into an understanding of the atom, and after all that has happened since, we find that we are a little like Christian Rosenkreutz and Frank Oppenheimer combined. And we are left with a question as to how we will wake Venus. Are we going to stand there before the open veil and ask how we can "get this thing to go" or are we going to gently wake her so she will respond in an affectionate manner that will give birth to kings, to a new consciousness that understands how to rule the earth with loving regard for the whole of nature? The choice is ours. But we might consider while making it what Venus might have to say about it when she wakes. After all, we could have a bomb in our hands.

3. Pierre Curie (1859–1906) was the son of a physician. Dedicated as a young man, he was home-schooled before entering the Sorbonne where his passion for science quickly took root. As a young man of 19 he became the lab assistant of the Faculté des Sciences while attending courses. After graduating in four years he was appointed chief of the laboratory at the School of Physics and Chemistry of the city of Paris. There he did the research on piezoelectricity and magnetism that was to be of great importance to the broadcasting and communications industries. He also published a seminal work on magnetism. (*The World of the Atom*, pp. 427–28)

4. Apparently there was still some uranium in the ore. In their writings on how they extracted the polonium and radium, the Curies speak of uranium and thorium remaining in solution. (*On a New Radioactive Substance Contained in Pitchblende,* by Pierre and Marie Curie, Comptes Rendus, 127, p. 176, as cited from *The World of the Atom,* p. 433.)

5. The following rough outline of the procedure is taken from *On a New Radioactive Subtance* by Pierre and Marie Curie. The directions in the original are not meant to be exact and are here paraphrased with the intention of giving some idea of the analytical challenges involved. It is hoped that this will also give an approximate idea of the procedure used. Essentially it involved the isolation of the sulfides of bismuth and of an unknown active substance that proved to be over a hundred times more radioactive than uranium. This method was initiated by treating solutions with hydrogen sulfide. The sulfides then obtained were then treated with ammonium sulfide which removed the sulfides of arsenic and antimony. The sulfides insoluble in ammonium sulfide were dissolved in acids to separate out as nitrates or sulfates, in particular lead sulfate. "The active substance present in solution with bismuth and copper is precipitated completely with ammonia [along with bismuth] which separates it from copper (?). Finally the active substance remains with bismuth." And then the admission that "we have not yet found any exact procedure of separating the active substance from the bismuth." One can sense the calm frustration mixed with determination. And then something of a breakthrough but expressed with the same steady calm. It was discovered that heating the pitchblende cause the sublimation of "some very active products." The story goes in their own words: "This observation led us to a separation process based on the difference in volatility between the active sulphide and bismuth sulphide. [For this purpose] the sulphides are heated in vacuum to about 700° in a tube of Bohemian glass. The active sulphide is deposited in the form of a black coating in those regions of the tube which are at 250° to 300°, while the bismuth sulphide stays in the hotter parts. (*On a New Radioactive Substance Contained in Pitchblende,* as cited in *World of the Atom,* pp. 433–4.)

6. Let's do the math on this. From the diagram of the tube above we have the two plates, D and E and an electron beam between them that is being deflected: The deflection and what determines it is the core of the problem.

This deflection, or the angle of deflection would be, for the purposes of the experiment, the same as the ratio of the velocity, V_1, in the direction of the discharge (along the x-axis) and the velocity, V_2, in the direction of the electrical field, F.

The velocity, V_1, in the direction of discharge is the unknown velocity that will be found at the outcome of the experiment. The velocity in the direction of F, or V_2, can be seen to be directly proportional to the intensity of the field, F, the charge, e, and the time, t, between the plates but inversely proportional to the inertial mass, m, of the charged particle. Since this exhausts the variables we have the equality,

$$V_2 = F(e/m)t \quad (1)$$

Thomson set the time equal to the length of the plates divided by the velocity, V_1:

$$V_2 = F(e/m)(l/V_1) \quad (2)$$

Since the angle of deflection, ø, has been set equal to V_2/V_1, we have

$$ø = F(e/m)(l/V_1^2) \quad (3)\ 1$$

If, instead of an electric intensity a magnetic force, H, acts on the charge particle, we have to allow for the horizontal velocity, V_1, to have a direct effect on the deflection. This is because any charge in motion creates a magnetic field just as the current in a coil of wire creates an electric magnet. The formula for V_2 with a magnetic force, H, thereby becomes

$$V_2 = H(e/m)V_1 t \quad (4)$$

and,

$$V_2 = H(e/m)V_1(l/V_1) \quad (5)$$

makes the angle of magnetic deflection, ø,

$$ø = H(e/m)(l/V_1) \quad (6)$$

Since V_1 is the only velocity we now consider it can simply be V. Dividing (6) by (3) we get

$$V = F/H \quad (7)$$

and by squaring (6) and dividing it by (3) we solve for m/e:

$$m/e = H^2 ø l / F ø^2$$

In the actual experiments, Thomson adjusted H so that ø = θ.

This let the equations become

$$V = F/H \text{ and } m/e = H^2 l / F θ$$

Since all of the values could be found from the apparatus the values for m/e and V could be determined. (Adapted from "Cathode Rays," by J. J. Thomson, *Philosophical Magazine*, 44 (1897), pp. 293–311 as cited in *World of the Atom*, pp. 422–23)

7. *The World of the Atom*, p. 419.

8. It is important to know here that these conclusions were found to apply to any gas and were therefore universal for gases. This is one of the main considerations that led to Thomson view that the electron is a universal particle. (*The World of the Atom*, p. 425.)

9. Ibid, p. 426.

10. Lockyer and Remsay get equal credit for discovering helium, though Ramsay is given credit for isolating it as and earthly element. This makes him the only scientist to have discovered an entire period.

11. In the French speaking part of the world credit is generally given to the French scientist, Paul Villard, for discovering gamma rays in 1900 But across the Channel, the ascertainment of their wave lengths by Rutherford and Andrade in 1914 quantified their existence and formal credit is generally given to Rutherford for being the bona fide English discoverer. He had spoken of them as early as 1903 and considered them a part of his atomic "family" though not on as intimate of terms as the α-particle. (*Rutherford and the Nature of the Atom*, p. 69.)

12. Changes that produced α particles were not understood till the discovery of the neutron in 1932.

13. With the discovery of the neutron it was understood that α particles must come from neutron decay. Such decay explains how Thorium, for example, transmutes into protactinium. When thorium-234 decays it shoots off an α particle, a high-speed electron from the nucleus. To make this happen a neutron decays and splits into an electron and a proton. The thorium nucleus has 90 protons. With the decay of a neutron and the ejection of a α particle, however, it has one more proton. This gives it 91 protons and turns it into a new element, protactinium.

14. An example of the particulate behavior of energy occurs whenever when take a photograph using black and white film. Photons of light energy initiate a reaction that reduces the silver halides on the film ($AgCl$ and $AgBr$) to yield free silver (the dark part of the negative). An example of the wave behavior of energy is in the refracton of light that produces rainbows or in diffraction phenomena such as the rainbow colorations of oil slicks in the gutter after the rain falls.

8
WHERE IT'S ALL GOING

So when all is bread and done, have we gone astray to claim that man doesn't live by toast alone? Or a plate full of crumbs? Or so many atoms in a bowl of alpha-pet soup? That is, in so many words, what Berthelot was asking in our "Interlude" as he lamented how the "entire material universe is claimed by science." But he might have drawn some consolation from the fact that the material universe belongs to nature and nature will not let us be complacent. To keep up, science has to stay on the move.

We've seen already how that goes; science never was still—it always was changing. Indeed, it's a little like those moving electrons that generate all those electromagnetic fields. In fact we are about to see science generate some new fields of its own. To get the field of that one, we will be focusing on some recent changes that are causing some rethinking and some reshaping of some old paradigms. Which, of course, is the way it should be. Thomas Kuhn dubbed such rethinking and resorting of established ideas a paradigm shift in his *The Structure of Scientific Revolutions*—and it is the purpose of this chapter to give some indications where the shift is shaking.

But be forewarned: when shaking up old ideas one can shake up the people who hold on to them. Some things can happen that are, well, revolutionary. That might not mean so much to you but for older folks (like me) there might be some cherished notions out there—notions that reflect how we once upon a time read (or misread) the book of nature. Fortunately, nature with her eternal play of process, could care less about that. Which is great if you like being natural. It keeps you moving and fit—and ready to adapt. And ready play a bit with the notion of survival . . .

Meanwhile, the notes of Eliot's violin still play among the many strings and threads that connect this universe and the struggle to understand it. So if we listen to the stillness of Eliot's Chinese jar, the stillness of a violin while the note lasts . . . can we hear where the song is taking us? Beneath the sometimes noisy weave of patterns, particle and wave, can we still hear the stillness? What's that? A ringing sound? Betchya it's our friend with the melodious voice calling to see if we made it to the meaning of science. Or at least to the end of the course. It is! She laughs—asks if I still like her marriage plans . . . yea! To the tune of that? Sure, with that kind of harmony to tie the knot we'll unlock the door to science with a major key! Got it . . . the beat . . . the cosmic beat . . . star dust in the blood . . . the rhythm . . . the proportions . . . no, no, can't do without 'em . . . everyplace we go, everyplace we goho! Hey, we'd better watch it. We have readers here who are trying go figure out what we're so goho about. Sure—it's all right there—from Proust's law of constant proportions to Gay Lussac's law of volumes and—you know, maybe it would help if they knew you were having a little party over there—complete with a chorus of voices doing the acappella with the periodicity of the elements and the predictability of chemical properties . . . and well . . . why not? With all the other bits and pieces of matter, the crumbs of substance have some music in them . . . that's no news. That's it . . . yeah . . . yeah . . . .oh yeah! With some cookies for Prout and a few pretzels with Crookes in them along with Thomson and Rutherford with their intimations of primordial substance, and the way the atomic weights frequently vary by steps of one, two, three or four . . . we've covered that! Hear all about it: hydrogen and its fat cousin, helium, the building blocks of matter, yep—yoou've sure got that one right. What's that? A change of tune? Put on something from those rapper guys who keep going on about that bunch of sunshine in their eyes—yeah, they're the ones. Ha! OK, if we make it with those fields of life in the shine, sure thing! You know, when you laugh like that it really makes me wonder . . . that laugh . . . sounds like it's got a fancy knot and some secret under the wrapper . . . what? Ah yeah—right. That's the cool part. Got it—yeah—the connections . . . I'm listening . . . . . . the connections between them are what really matters, are what really tell us what's more than the matter with matter . . . Right! Relationships—that's the word. That's what gets the chemistry flowing, what takes the solo out of the goho and gets to where it's all go . . . What's that? You want to leave us with a question to help us get there? How Waldorf of you! Yeah, yeah, fire away. Hey! Whoa! Not so fast. Let me translate in slow words for our reader: She started rapping with her chin on the spin about how we need to sow the flow of the know where the jar's gotta goho so we can begin again with the trend on the mend . . . OK, I missed the question part but you'll have to admit I was pretty close . . . ha! I like that. Sounds like were on the same note to me . . . don't worry, we're looking out for that one! Yeah? Where the jar is on empty because it's so full? You got it!" Hey, don't . . . go . . . She's gone. And I still didn't get her number! Oh well. What was I saying? Oh yes... "Where it's all going . . . " Good question . . . and come to think on it, what *did* she mean by that "jar on empty because it's so full" bit . . .

Ah, what does it all mean, that is the question . . . Well, for starters, I think she means to ask if we're going to spend the rest of eternity spinning around that Chinese jar. And that's a big one. But we have already answered it at least half way. In our little story of chemistry we have put human beings—and above all, ourselves—inside the jar. That was a definite step in the right direction. It centers us in a way that lets us see all of the jar rather than just our side of it. It enables us to gain a holistic view of how the whole thing turns, of how the past relates to the present and even to the future. And this gives us an indication of where it's all going. See how simple that was?

Rinnng! Uh oh, sounds like our melodious caller again. Halloha to you too! We thought it might be you. Thanks. She hates to disillusion us but being simple is not where it's at. That's not where it's all going . . . What? Complexity . . . yeah, of course I've heard of it. You don't think I was going to Simple Simon our way out of here . . . No . . . no . . . right! We don't want to follow those Chinese dragons . . . yeah yeah, the ones with the theoretical green fire in their eyes and the brown spots on their tail—those are the ones . . . . .uh hum . . . .reducing the story of the world to a flip of the molecular tale . . . You know that's not what I meant by . . . simple! All right, we'll look out for those kind—they can chase their tales till the world spins dry. Right! The safest place to avoid them being in the heart of the jar . . . right . . . where the old turn ends and the new begins . . . sounds like you've been there! Thanks! Right . . . where all the pieces turn together . . . along with the other still points of the turning world. Gotchya. Give you a call when we get there . . . She congratulates us for the little bit about the holistic view—put her own tap on the rap about taking the solo from the goho . . . And chimes off with a melodious hint as to where the next part of our answer might be.

Ah! Don't say it! Did it again. . . Oh well, I guess as long as she has *our* number when we get there . . . Right now, it's the getting there that counts. And we do that with a melodious hint . . . and if it gets complex, well, that's life. Soooo . . . Ms. Melodious, here we go. Melodious. Sounds like we need to think back again, back to the energy dance as we science along with the proton and the electron and hit the beat of the elemental band and discover **resonance**. Just don't, with all the musical metaphor, get any ideas that resonance has anything to do with how your vibes jive. Here the word resonance has to do with a dance of a different color. We can spare ourselves the technical definition and paraphrase Linus Pauling, the Nobel winning chemist who came up with the perception of how it happens. Resonance, as we speak of it here, happens when an electron bilocates or even trilocates, and finds itself in two or more places at once (my words, not Pauling's). Or, to be more exact, it delocalizes and acts experimentally like its covering a multitude of bases. The important thing is that with resonance we have another example of how electrons are always *moving*. In fact, if we want to keep in party mode we might think of resonance as describing a kind of electron dance as electrons do a kind of ring around the molecule.[1] But the point is this (and I say it again): the electrons are moving. And with a little recall, we remember what Michael Faraday said about moving charges—that they create magnetic fields. Fields. As you recall, we mentioned fields with regard to the ions of Arrhenius. I won't go over it again but if you've forgotten it's till back there where we left it—something about ions being another pattern in the jar, as I recall. Or rather, another pattern in the field. Anyway, if we keep fields of energy in mind it's all a matter of learning to read the patterns on the jar so it's the jar that spins, not our heads. The moral of that tale is this: play the field. Give the patterns and symbols some room, some space for action at a distance, some spatial swing, and they become dynamic partners in the dance. And as we dance we can re-interpret the terms polar bonding, co-valent bonding, etc. The one holds the jar together, gripped in field of energy, the other enables an organism to create the myriad forms and functions that make life a vast web of interacting fields. For as any dragon can tell you, what happens between atoms in any compound, organic or not, represents some sort of field/energy exchange, some sort of giving and receiving of energy that causes the electron energy—the field energy—that surrounds an atom or molecule to connect in predictable ways.

With people the energy fields connect when two or more individuals are on the same beat, when they warm up to each other and discover how the *quality* of relationships is no simple affaire. With the molecules it's a little less personal. But you probably catch the drift—we're getting warmed up to that notion of complexity. And with that notion of complexity, to how reactions depend on the nature and the quality of the fields that make them happen. So I want to say a few words about quality.

If we get specific about what we mean when we refer to the quality of a substance or substances, of events or phenomena, it will help us cover the ground ahead. Now it's rather obvious from what we know so far that some substances are more conducive to exchanging energy than others. We have seen this with regard to various elements (nitrogen, phosphorus, etc.) and compounds (ATP, nucleic acids, proteins, etc.). For now however, to show how complexity can work on a nuts and bolts level, let us consider the titanium screw. Used in surgery when it is desirable for the metal to bond with bone, the titanium screw is a screw of exceptional quality. Stainless steel used to be used for this but it caused complications due to stress corrosion and cracking. Stainless steel is also subject to pit corrosion and consequent galvanic effects that make the situation worse, especially around bodily chlorides like common salt. Titanium, on the other hand, has none of these defects. The two metals, one an element and one an alloy, have different properties, different crystalline structures and different electron configurations. Be this as it may, it is likely that several properties of titanium are responsible for its unique relation to bone growth. These can include density, specific heat, valence, whether it is magnetic or paramagnetic or a good or poor conductor and so on. It would be interesting, for example, to experiment with how electric fields affect bone growth and determine if titanium somehow contributes to the process in a way that was different from stainless steel.[2]

As with all organic processes, force fields play a contributing role in how things grow. But in any case, no experiment can eliminate all other properties of the metal and isolate one single property responsible for the unique relation to organic process any more than we can eliminate all players on a basketball team but one and determine if that one is responsible for winning the game. This is because organic process entertains a complex array of reactions (the team) that requires a unique set up (s/he shoots . . .) that depends on how several proteins (the players) act together in response to several metallic properties (the relief players) in a comprehensive manner (that includes the whole team—hey, get that rebound!). This team character of several properties working together makes it necessary to use the word *quality* when referring to an effective relationship that wins a ballgame.[3]

And so it goes from titanium to teamwork. To go from there to chemistry, consider how the magnesium in chlorophyll gets changed to iron to form a simple hemoglobin—a transition we've discussed on numerous occasions with regard to how an array of interactions and relationships depend on molecular structure and elemental properties working together. Teamwork scores again. Or in the way we use catalysts in the lab to accomplish what enzymes in an organism achieve with such admirable efficiency. Score again. Then, for some teamwork of really exceptional quality, there's always the great example of rhizobium bacteria and legumes and how they join forces to form an oxygen carrying leghemoglobin so the nitrogen fixation process can fix nitrogen with oxygen and create nitrates that contribute to the growth of more chlorophyll, proteins, enzymes and DNA. Such symbiotic teamwork gets complex (there's that word again) as it creates an ongoing cycle of mutual dependency. But such relationships are common and show how life depends on organisms working together as a whole—and the qualitative nature of life determines how well this works. In fact the more we look for team action in nature the more we seem to find a qualitative array of relationships and properties in full court press. It all shows us how the qualitative and the quantitative can express valid aspects of a process that is fundamentally holistic. Like an energy field. Or like a symphony. Stay tuned.

But first a little word form our sponsor. Breath. Never go anyplace without it. Take a deep one. And exhale. Or whistle a little tune. That's it. We take a deep breath and admit that there really is some lively music there. But we can't stop—no, please, whatever you do, don't stop—we really must keep breathing and *moving*. After all, we are on a quest, no time to rest, you know how it goes. So let's go a little further into the past to see our way to the future. A little further, back to Priestley's idea of breath and all that it took to raise Adam from the dust. If we really dust off the past, we even find that Priestley was not alone in having such a lively regard for breath. Wolfgang von Goethe, whom you perhaps remember from when we studied the archetypal plant in the ninth grade, also saw breath as something of an archetype. For Goethe, breath was what made the dynamics of nature possible. In more modern terms, Goethe's "breath" was a metaphor for the giving and receiving of energy, for the inhaling, exhaling, expanding and contracting of all that unites us with the in and out stream of co-existing fields and forms.

In many ways Goethe reminds us of the archetypal and universal thinking of Greek philosophers, but in other ways he was every inch a modern scientist. He was a very astute observer of natural phenomena and felt that his work in science was his crowning achievement, though he was to attain far greater recognition as a man of letters. But this may be changing. It seems that part of the world might be about to catch up with Goethe. As a thinker and observer he had little use for the analytical mind that reduced natural phenomena to a mechanistic model. He was concerned with how nature worked as a whole, with how the bones of a body shared a common formative principle, with how the leaves of a plant metamorphosed from a parent form—how the parts related to a greater whole. When he observed nature he saw systems of dynamic process interacting with other systems. Though he didn't quite say it this way, for Goethe nature was a vast array of interconnected and interdependent relationships. In seeing natural phenomena this way, he anticipated some modern trends of thinking that are at long last challenging some of the old dragons who might be a trifle too fond of their place on the jar—so fond that they cling to some old paradigms and theories without seeing that the jar is turning beneath their claws. So take a little advice from our sponsor. Think of Priestley and Goethe and all the other denizens of the breathing world. And take a deep breath. Take quite a few in fact and take some along for the next step of our journey.

By going backwards to find some thinking that was many years ahead of its time, we are making progress in our quest. The sound of the violin is in our ears . . . playing some grand old theme that leads us back to where we are . . . and to where we are going . . .

Nature is as full, it seems, of surprises as we are full of ways to see her. She loves chaos but relies on highly complex systems of intricate relationships in order to keep her organisms alive. Recent breakthroughs in biochemistry provide a vision of these relationships and in so doing open windows to some much needed fresh air. So take another of those deep Goethean breaths. We have arrived. Take a giga whiff. Smell it? Complexity is in the air. And since this is a book on chemistry, we might as well start with what a biochemist has to say about it. Enter Micheal Behe. He has taken a very close look at the biochemistry of life, at what the molecules do in living organisms, and at the complexity of molecular interactions that make life possible.

The result of such work done by Behe and numerous others is a new science of complexity that crosses over into many fields, connecting biology and physics with biochemistry right in there wielding its tools of technological precision. The cutting edge of these tools is a sharp one. It allows us an almost intimate look at what the molecules are doing and how they interact. The use of computer models to cross check data gained from living tissue enables scientists to check interpretations of life processes without interrupting them. Or if they do, the results can be integrated into a view of how an organic process works as a whole. But even with the gadgets, the word interpret is a big one. To interpret and correctly account for what one sees requires a perceptive eye and mind coordination and a healthy dose of honesty.

Behe combines humble metaphor, a few jokes and a lot of science to qualify. He writes with flowing ease, combining technicality and humor with respect for the facts of biochemistry. But it still comes as a surprise when he compares nature's complex molecular mechanisms to a Rube Goldberg device—the kind where a lever is timed to drop a ball into bucket of water to splash an inclined plane and cause it to tilt and free a waiting wheel that strikes the end of a hammer that hits a nail that pops a balloon that wakes up a sleeping daredevil canary that jumps off its perch to land on the back of a snoozing cat who swats at the laughing canary and knocks over a glass of water caught by a funnel that drips onto the forehead of a sleeper and causes him (or her) to wake up and put the cat outdoors: that kind of device—the irreducible kind where each part has to be there to make it work. If the canary doesn't wake the cat misses a night on the prowl and the canary gets a night with the cat. Not good, even for daredevil canaries. Fortunately for our canary, every aspect of the intricate awakening device was irreducibly in place, every step did its essential bit to make the whole thing work. Behe uses such Goldberg cat and canary devices to parody the irreducible complexity of, say, the intricate interrelationships and reactions that cause blood to clot. Bringing in the science, he shows how the intricate processes of blood clotting work down to the last protein interaction. What emerges is a detailed exposition of a cascade of reactions that all have to be at the reaction site (the wound) and ready to react. If any single step in the whole series of the reactions is missing, the whole system fails. Blood doesn't clot and the wounded animal (or human being) bleeds to death. And that is only one system.

One among thousands. And all the systems depend on one another just like all the parts of each system depend on one another. Small wonder then that organic complexity forces us to see the body as a whole organism determined *not by its parts but by how the whole system works together as a mutually integrated unit.*

And here we get to the crux of the matter—and to the controversial part. I did warn you about as much at the start of this chapter. So here goes. The controversy revolves around what we have been taught to think about genes and their role in evolution. The standard paradigm of neo-Darwinian evolution theory says basically that random mutations cause changes in genetic structure that can create some novelty in an organism that makes it better able to survive. That process is called natural selection. Behe's findings give rise to many questions about how this process might happen and his ideas challenge some cherished models of current theory. But wait, you might say, isn't this a book on chemistry? Yes, indeed—and that is precisely the point. It is biochemistry to the tune of complexity theory that is raising so many questions and providing answers that raise even more questions. In other words, biochem is raising a ruckus. And who wants to miss a good ruckus? Especially when it happens in our backyard. So here we go.

Behe entitled his book *Darwin's Black Box* for a reason. Complexity is like a dark voice in the closet of evolutionary theory and neo-Darwinians would just as soon it stayed there. But out of the closet it comes—such things never stay in there once they are known—and the doctrine that says random mutations are the prime movers of evolution is in for a shake-up, to say the least. Let's see why. Let's say an organism gets lucky and against some very stiff odds a random mutation produces *one* of the protein agents necessary for the blood clotting process (one among thirty or so). But the organism also needs to produce a trigger protein that activates the blood clotting agent. Otherwise, the blood clotting agent would cause all the blood to clot and the animal dies an excruciatingly painful death. No organism wants that to happen. So let's say our happy organism gets *very* lucky and another "random" mutation gets that step right. Now it has a protein blood clotting agent and activator ready to go. Bravo! Bully for our organism. But now the question arises, go where?

The protein agent-trigger duo is useless without the other twenty-nine or so proteins necessary for the whole process. And for *random* mutations to produce all twenty nine in assembly line order would be something of a contradiction of terms. According to Darwinian rules, it ain't gonna happen. So what does an organism do with useless proteins? Well if we follow the ground rules of natural selection any novelty (read random mutation) that doesn't contribute to the immediate survival of the system is like excess baggage—it either gets tossed or sent to the recycle bin. Don't need that stuff on the steep stairs of life. Blood clotting? I'll get by with a system of vascular shut off valves, thank you very much. (But we won't begin to go into the complexity of that one.) Or, if things aren't so simple—and recent research suggests they are not—and a novelty, say, gets "put on standby," then we have to explain how that can happen with yet another complex set of relationships. And in this case, it's a set of relationships that only has a reason to be in the future! (Which reminds me, would anyone like to donate a few precious moments to the cause of our science and do a report on junk DNA? You will find that it is a good example of seemingly superfluous substance that relates to either the past or future organism or both.)

So it seems that regardless how you cut it complexity is where it's flowing. (Did I say cut? Ouch! Oh I do hope that prothrombin and thrombin aren't making out in the corner somewhere when I need them to get my blood to clot!) Yes Matilda, simple solutions are out; complexity, it seems, is in. But if you are about to cut in (Ouch again!) and object, let us hasten to add that none of this in no way contradicts evolution, it only demands that we rethink some notions about how it works. The ground rules of evolution might be shifting, but evolution is still on the map. Which is to say that Behe's findings have caused some rumbling among the orthodox map makers, among those who adhere to a more orthodox neo-Darwinian view. Darwin and the neo-Darwin camp placed all of their bets on the viability of random mutation—on a change of part, not to be confused with change of heart—and how it leads to natural selection. If that doesn't work, the theory bites the dust. Or, like the primal Adam, it's in need of some of that breath to get going again. But these days that breath is coming from the biochem labs. And that is why complexity has given our Chinese jar a new spin. The patterns on the jar are changing. Even if it means that we have to break the jar.[4]

One would hardly want to call Behe a jar-breaker. He is a scientist who looks eye to eye with what he sees and gives an honest interpretation. Just like scientists are supposed to do. In this case, however, the honest but inevitable interpretation returns from the past like another pattern of our ever-present jar. The pattern is called intelligent design and it harkens back to the Greeks and beyond when it was felt that man was built in accord with the divine image, etc. Like a lot of old notions the concept of intelligent design got some fresh wind in the seventeenth and eighteenth century but was pretty much booted out in the nineteenth by natural selection and Darwin's *Origin of the Species*. But now it's back in the controversial soup of science as another part of the process. And it has some folks stewing. Behe revives the topic of intelligent design because as a scientist he has to confront the Rube Goldberg nature of organic reactions that seem to defy any random origin. They reflect intelligence in the intricacy of their interactions; they reveal the presence of Rube—of conscious design. And for most scientific folks there's the rub: the words intelligent design are too close to Rube for comfort. Intelligent design is not something one should say in the polite company of the scientific minded (not to mention the materialistically inclined). That is because, as Behe freely admits, on the other side of that almost four letter expletive resides the implication that there might be some divine author of this intelligent design. (Egad! A divine author with the name of Goldberg! Heaven help us . . .)

But of course you see the ridiculousness of the situation. You see that this pits something subjective and irrational against a long tradition of empirical reason that demands a strictly objective world. Conventional wisdom has it that if you throw a divine factor into an equation, God knows what will come out. Even if pure chance takes it from there, the odds are pretty good that it won't be science. But Behe is not conventional and yet he is very much a scientist. In the spirit of science he debates the pros and cons of the issue, exploring different views and essentially concludes, as numerous other scientists have done before him, that "religious" views, whether philosophical or a professed belief, are personal matters and should at all costs be left out of science. Science has its own guidelines and procedures that define its range and method of inquiry and these need to be respected if the word science is to have any meaning. This is not to say, however, that science is not about the expanding of consciousness through observation of natural phenomena. And of course, there are no limits to this expansion. The mind must above all be free. In a concluding section entitled "Don't Fence Me In" he states that "a man or woman must be free to search for the good, the true and the beautiful."[5]

One might add that a lot of Waldorf students might agree. They might even add that with regard to fences it is the quality of thinking that opens the gate of seeing to a vision of relationships that is free of philosophical limitations. To achieve this, science demands a certain professional honesty that prevents a scientist from adulterating his or her science with a personal agenda. With this in mind Behe gives the knife a little twist. He is very concerned with the quality of thinking that goes into science and like other scientists very disturbed when religion or philosophical bias does creep into science via the back door. In fact he warns against this very thing regarding how some tenants in the tree of evolution cling to their theory as though it were something of a religion.[6]

Behe is saying that we can stand outside this tree and see it in a new light. In this light—the light of complexity—we see how organic systems work as a whole to include genes as partners in the life of an organism. This, as we will see, has far reaching implications that go well beyond any controversy over whether complexity is intelligent or not. In fact, we might even say that for the sake of our present argument, the topic of intelligent design, though full of the spicy stuff that makes the story of science vital and interesting, is just another pattern on our jar. The bottom line is that Behe's work suggests several new ways of looking at chemistry and biology, ways that can prove seminal if we don't become bedazzled by those green flashing dragon eyes that end up chasing those spotted tails again and again . . . with nary a thought as to where they lead or where they end . . . and all the while jarring the world with spinning theory to explain why the thing is turning so . . .

All right. Those were my words, not Behe's. But regardless whose words they are, those spotted tails somehow remind me of Paracelsus.

Only the Paracelsus of old wouldn't have said it so nicely. He was a man who called a dragon a dragon and who knew dragon dung when he saw it. Fortunately, to spare us the smell of dragon dung or a pile of feces on the lab table, we have some modern versions of Paracelsian verve with a smoother flair for words that flows with a cooler nerve. Brian Goodwin, author of *How the Leopard Changed its Spots*, takes us right to the heart of the jar. In the introduction to his book he writes that

> The recognition of the fundamental nature of organisms, connecting *directly with our own natures as irreducible beings*, has significant consequences regarding our attitude to the living realm.[7] (Italics added.)

Now when anyone says that something has significant consequences you know that person is serious. And since we are a part of the living realm it is kind of nice to have someone take us seriously for a change. It's the kind of attitude we might like to see more of in the world. But what on earth does he mean by the "our own natures as irreducible beings" part? As with Behe above, the word irreducible is thrown in to mean that we are defined by our complexity not by our parts, that the bright smile we make in the morning mirror can't be reduced to a pile of molecules, that the whole really is greater than its itty-bitty pieces. In this way you are like a mousetrap. (Sorry, I just saw that big smile and I thought of cheese.)[8]

Anyway, mousetrap. We can't define a mousetrap by its spring, by its snapper, by its cheese. If we take away any of its parts it won't work. It's irreducible. Take away the spring, the snapper, the cheese and the thing is useless. So you can relax and smile some more: your smile expresses a *quality* that is uniquely yours as it represents the *whole* you. With smile you remain, bright and wonderful, a holistic unit. Likewise, Goodwin adds that an organism functions as a holistic unit. In so many words this means that our molecules, our genes, our choices, our personal history, social environment, the works—all of it works together to make a mutual contribution to who we are. Irreducibly yours truly, we are more than our molecules. And with all of that his book asks us to take off the tinted glasses of any theories or philosophical attitudes that predispose us to think a certain way and see with eyes that approach an organism, not as a complex molecular system determined by genes, by genocentric biology, but as a whole that functions as an organocentric unit that includes both the organism and the space in which it lives. We might recognize this holistic approach as sounding somewhat familiar, being as it is so intimately a part of Goethean thought. So take another of those deep breaths. Goodwin writes that "Goethe believed in a science of wholes—the whole plant, the whole organism, or the whole circle of colors in his theory of color experience. But he also believed that these wholes were intrinsically dynamic, undergoing transformation—in accordance with laws or principles, not arbitrarily. So he was an organocentric biologist, and a dynamic one to boot!"[9] Goodwin concludes that "the ideas I am developing in this book are very much in the Goethean spirit."[10] So you see how this last chapter of ours follows a very definite theme? It's taking us somewhere as quick as you can say the word complexity. So onwards! We need to look more closely at what it means to be "organocentric" and how this will take along the way on our quest to see where it's all going.

Remember Descartes? Remember the mechanical model? Remember the Cartesian partition? Good! Now we can take the ball and play with it. Goodwin begins the last chapter of his book with the question, "Is an organism a mechanism?"[11]

Swish—right in the net. There it is. Right to the heart of the matter. On the scoreboard, his answer will sound familiar. He sees organisms as holistic expressions of what he calls a "morphogenetic field." (You caught that word, "field," right?) Morphogenetc field is a concept he borrows from physics as much as from biology and it refers to the organization of space about a living organism in much the same way the magnetic field organizes space around a magnet. If you inhale really deep you can catch a whiff of chemistry as well. Look at that ball—it has resonance written all over it as you pass it around so fast it looks like its two places at once. The action becomes ever more charged with a field of energy that pushes the game to the point where—yeah! Score! (Deep breath.) Now take this analogy into the realm of biology and similar factors score a morphogenetic field but do not stop the game to define it. A morphogenetic field is defined by its own dynamics. The key is movement: a complexity of moving parts energizes an organism with a field of its own. The terms are borrowed from physics but applied to biology. By borrowing from physics Goodwin is doing what we are doing with basketball: grafting his biology onto a viable set of metaphors and relationships. He is not saying that biology can be reduced to physical laws or to a mechanical model. He is very clear that an organism is a self-ordering entity.

Unlike a machine that is assembled from premade parts, an organism creates itself. In case there is any doubt he adds that "organisms are not molecular machines; they are functional and structural unities resulting from a self-organizing, self-generating dynamic."[12] It is this self-organizing and self-generating game plan where all the parts work as a whole, as an irreducible team, that makes an organism a dynamic entity, that gets it going as a morphogenetic field and distinguishes different species. In this light a house cat and a mountain lion have different morphogenetic fields; so do a dog and a hyena. Each field expresses its nature "*through particular qualities of form in space and time.*"[13] And there's that word **quality** again. You can probably guess that it refers to a dynamic collection of traits that cannot be measured or quantified. Goodwin stipulates that it is the dynamics of this collection of traits that distinguishes different species (house cat from cheetah, etc.).

And here is where the organocentric part comes in. As part of the overall dynamics of the organism, the genes do their part as regulatory factors in the reproduction of these traits. But they do not determine how these traits came into existence. The origin of what makes a house cat a house cat and a cheetah a cheetah is found, according to Goodwin, in how the morphogenetic field of a whole organism achieves optimal fitness with a minimal expense of energy. How this happens can entertain a complex array of relationships between an organism and its environment, relationships that bring into play a whole host of internal response factors such as reproductive habits, diet, choice of habitat, climate adaptation, etc. All of these factors are expressions of nature that achieve "particular qualities of form in space and time."

To see what this means on the human level we might play a bit with how you or a friend might look on any given day in the qualitative history of your life. You dye your hair red, wear shades and the latest in designer jeans that look like they barely survived Omaha beach (with tank tops to match) and you have a certain *quality* that would not be there if any of these necessary glad rags were not a part of your attire. There is no way I can measure this quality; I can only observe and note that it is there. The measurements you add to that (which may or may not add to your self esteem) would naturally include your quantifiable properties such as your electric charge, magnetic force, heat content, weight, mass, position, or even the notes you sing on the way to chem. class. But you are not just singing just because you like chemistry—no, you are singing because your song is more than its notes. You know this because you feel the music you sing. And you know that the movement of a tune is a series of musical gestures that passes through variations on a theme and that it must have a certain *je ne sais quoi* that makes it *real*, a certain quality for it to knock 'em over and shatter wineglasses. And with this realization (which, of course, you got from not being late to chem class) you become an operatic success and a happy diva. And in your next lifetime you become a chemist with holistic vision or a perhaps a biologist with a Goethean tune.[14] You find it natural to compare organisms to music; you discover how musical qualities pertain to "real" aspects of an organism. Whew! And after all of that you understand how these qualities form an organocentric whole as the mind's eye learns to integrate living functions into a vision of self-supporting relationships so that every bone, every corpuscle, every muscle, every ligament are like one huge extended Chinese family. And you have this vision of a Chinese jar, turning, turning, and among the patterns you see from the heart of the jar you see life organocentrically. You see how all the parts of an organism, genes included, work together to create an irreducible whole. Like a song, like a tune that might begin when someone turns a chaotic feeling into a note that leads to another note… And you understand what the phrase "emergent order" might mean and you can see how it might lead to some very complex music in the realm of life. You know because you have just lived through a very analogous series of events . . . And the chances are pretty good that you aren't even Chinese.

On a slightly different note, the natural consequence of all of this comes to what Goodwin calls a science of qualities that recognizes how important relationships are on all levels. You have probably thought along these lines yourself after watching the latest T. V. news. Just imagine, a world that defined culture not in terms of money and power but in the quality of relationships . . . That could mean that some so-called primitive cultures could be considered more advanced than some so-called advanced cultures!

And yet, biology depends on a harmonious quality of relationships all the time. Think what would happen if your white blood cells decided they didn't like the "ethnic profile" of your red blood cells. You might call that an attitude (or worse). You might think how unscientific (to put it mildly). And yes, you'd be right on both counts. On the other hand, what sometimes qualifies as science can harbor cultural attitudes that are almost that ridiculous. Like Behe above, Goodwin complains that a good deal of bad science has been shaped by social conventions that affect our way of seeing. (Our wag chimes here with the comment that this amounts to "insidious design.") We have, for example, the authoritarian attitude that infiltrates so much of Western civilization with the notion that someone or something has to be in control.

This is the attitude, he goes on to say, that has colored our thinking with ideas of genetic determinism, genes in control or genes as the "authority" that guides the rest of the body just as a autocratic parent or religion "guides" the unquestioning mind with dehumanizing authority. He and many others are saying that such concepts have little to do with science and a lot to do with cultural myths. The reality is that nature is not "red in tooth and claw"—a nineteenth century phrase that said more about human attitudes during the Industrial Revolution and growth of empire than about nature. Instead it has been found that the contrary is true—that nature depends on levels of cooperation that extend from the molecular to the cellular to the way eco-systems nourish a wide variety of species.

We played with this notion above with reference to taking "the solo from the goho" but such notions have become serious science. Fostered by the work of James Lovelock and Lynn Margulis and their

Gaia hypothesis,[15] ideas of cooperation have gathered a great deal of momentum since Lovelock wrote his seminal book *Gaia: A New Look at Life on Earth* in 1979. Goodwin advances this stream of awareness to a new level with his concepts of the morphogenetic field and collective relationships that foster survival. But if the quality of these relationships is impaired for some reason, be it natural or manmade, new forms or systems must evolve to meet the crisis. A science of qualities as Goodwin proposes, will help understand the dynamics of how this can happen.

Which begs the question, "Where does this lead evolution?" To answer this one we need to go back into the past again and work forward. Back to the wonderful sixties, so full of evolutionary fervor of a different kind. Not everyone was smoking things and listening to the Beatles, however. It turns out that a small group of biologists were keen on attaining a complete description of the tiny roundworm with a big name of Caenorhabditus elegans or c. elegans for short. Twenty years later the task was done: every cell, every neuron, every step in its growth process through its tumultuous teenage years and toward a not so orderly adulthood was carefully described and accounted for. All similarities to human evolution aside, the results were rather astounding. One of the researchers who led this effort (who, incidentally, locked himself in a room for two years to count cells) reported that "neurons are produced neither clonally nor from an orderly series of repeated cell divisions. Rather, they are generated by patterns that are 'unpredictably complex.'"[16]

The same researcher also commented on how "cell determination is autonomous; in most cases a cell is what it is principally because of its history rather than as a result of interaction with other cells."[17] And the most astounding thing of all: "about twenty percent die almost as soon as they are formed: these deaths appear to be intrinsically programmed suicides."[18] Now wait a minute. Is this a sensible way to run a business? What if you were running a business and 20 (twenty!) percent of your employees suddenly up and committed suicide? For no apparent reason! Why, not to mention what you'd feel, you'd have every lawyer in the country on your back! But with c. elegans, this is business as usual and typifies cell growth in general. One of the explanations given by the researchers was that such suicides often happen after sister cells that are really needed by an organism spit off. So we read on and find where in the case of neurons "cells of a different type can be produced by entirely different lineages in seemingly illogical ways."[19] The researchers were confronted with what they referred to as a "bizarre symmetry." Maybe, as one researcher observed with blithe admiration, "it's just the way the cookie crumbles."[20] Concluding comments by the leader of the research team recognized that neural "development is unlikely to be the result of a discrete, sequential developmental 'program' but instead is the outcome of a more holistic logic of molecular assembly."[21]

Well, you say, so much for how the cookie crumbles. But listen to another parting comment. Asked if knowing everything about C. elegans reveals anything about the rest of the biological world, one of the researchers replied that "although certain detailed aspects of molecular genetics have turned out to differ, many basic principles have proved to be universal." [22] So there you have it, ask not for whom the fortune cookie crumbles, it crumbles for you . . . However, before you blame the chaos or "bizarre symmetry" of your cells for not getting your work in on time, we need to ask how the crumbs get to be a cookie. After all, there might be a higher ordering power than your poor overworked cells. And here's your chance to prove it: if you've been an attentive reader the answer to what that higher ordering power might be should be rolling off your tongue. Need a hint? It has something to do with morphogenetic fields, right? Right! That's at least a big part of the answer. But you'll also be glad to hear that Goodwin adds another dimension or two. Along with emergent order we have life at the edge of chaos. (Yes, Matilda, you've got it! Emergent order! After the grand old age of 18 . . .) So let's hear it for the basic urge, the primal impulse at the edge of chaos. Something to think of the next time you and your skateboard take the plunge off the edge of that six-foot embankment. If you don't land so well, relax. Let Mother Nature show you how it's done.

Something similar happens when an organism evolves big time. One of the big problems with evolution as it's now understood is that small scale changes that have been verified don't add up to big changes in species. Birds and butterflies, for example, can change color to better suit their environment, but they remain birds and butterflies. No matter how good the wag, once a dog always a dog. Sharks don't turn into amphibious salamanders, no matter how much they want to take a stroll. Or, as Goodwin declares, to compete with existing organisms:

> Competition has no special status in biological dynamics where what is important is the pattern of relationships and interactions that exist and how they contribute to the behavior of the system as an integrated whole. The problem of origins requires an understanding of how new levels of order emerge from complex patterns of interaction and what the properties of these emergent structures are in terms of their robustness to perturbation and their capacity for self-maintenance.[23]

In other words, we need what is called higher levels of emergence to turn chaos into an evolutionary door of opportunity (if you don't believe me, read your fortune cookie). This is true all the way to the genetic level. The role of genes in the emergence is secondary. In doing the adaptive thing, "they cooperate with the generic forms of the (morphogenetic) field to give robust morphologies to organisms."[24] As Goodwin demonstrates many times over, genes don't wag the whole organism; it's the other way around. In keeping with our metaphors, genes are partners in the dance of life; they don't create the dance. Regardless how you shake it, the need to shake things up and resort the cells is what gets a new form to emerge, a new species to evolve. Based on observations of current life systems, Goodwin and others are coming to see life as existing at the edge of chaos, "moving from chaos into order and back again in perpetual exploration of emergent order."[25]

The researchers of c. elegans did not, to be sure, observe the emergence of a new species. They only observe a normal process of growth that perpetuated the same old show. But this same old show revealed an emergence of order out of chaos. They called it a "holistic logic of molecular assembly." What a quaint way to describe a morphogenetic field! But that was in 1984, some ten years ahead of Goodwin. The research group that explored c. elegans only reported what they saw (which is what they should report). Goodwin adds to this out of his own research on the evolution of form and finds how the pattern is still fit. The old pattern of emerging order, given the right novelty of the moment, creates a new pattern in the field of form and evolution takes one more step along the way. It is important to note, however, that no one can say what this new novelty will be, or what its specific causes are. This is determined by the dynamics of the organism and a complex array of other factors. Consequently, the science of evolution must necessarily proceed after the fact. But we have a new way of interpreting the appearances. We have a new pattern on our jar that is starting to go places. And from this we can gain some fresh perspective on where it's all going. From inside the jar . . . [26]

What this whole story from beginning to end is about, it seems to me, is in learning to see. And to do this we need questions that break the spell of old patterns, ideas that lead us on a quest for new ways of connecting with what we observe. Ideas that break away from the old myths of survival of the fittest, struggle after the fall, etc. Ideas that focus on the quality of life instead of the quantity of what we sometimes call life. And participate in the play of emerging new order.

To illustrate what this might mean to chemistry, Seyhan Ege, professor of organic chemistry at the University of Michigan, tells the story of a colleague who was becoming very frustrated with her experiments with corn DNA. Nothing crystallized right, the electron microscope might as well have been a Kodak camera, and if it were it probably wouldn't have worked either. So the woman just said in so many scientific words to hell with it and went out to spend the summer working on a farm that belonged to a friend—friends with farms are always in need of a few extra hands. So on the farm it was. Science out of mind and life in her hands, she went among the vegetables and yes, the corn and the tomatoes and the broccoli; she worked the soil, hoed weeds and in the dirt of it all got to know a corn she never knew. When she returned she discovered that everything fell into place. She got her paper written and I'm only sorry I don't have a copy of it. But the story behind it was better. One that that has a handle on it—a hoe handle, to be exact—right there in the still point at heart of the jar.[27] I wonder if she sang while she hoed.

If she did, it might have been some ditty about how chemistry and biology can sing a tune of exceptional quality. We might even sing along if we head out to our own garden, check out the sunflowers and try our own luck at finding some new ideas. And why not? Why keep our bumper crop of *helianthus argophyllus* waiting? Hi ho to the garden we go. With a little looking we find once again how the leaves spiral around the central stem with Fibonacci ratios of five leaves to two turns around the stalk, etc. We've been there and done that and you probably recall that we wrote down our leaf around the stalk observations under the rubric of phyllotactic patterns. You might even recall how we found different phyllotactic patterns in the same species, with some plants staying with the 5/2 leaf to turn ratio all along the main stem while others have a 3/2 toward the top. And sure enough, on this trip to the garden we find the same variation on a theme by Fibonacci. And we remember that over 80% of the 250,000 or so of the higher plants play a similar variation.

So what's going on? Too many patterns on the same theme to be random. So is natural selection stuck in a rut? Or could another factor be involved in deciding the growth patterns? To find out we read a bit and find where Goodwin ascribes this to the non-genetic activity of (you guessed it) a "robust morphogenetic process." To demonstrate how this works he refers to an experiment ran by two French scientists, Stéphane Douady and Yves Couder, who used a magnetic field to essentially duplicate the effects of a morphogenetic field.[28] The results of the experiment showed that all phyllotactic patterns could be generated by the magnetic field, the different patterns being a function of speed of plant and leaf growth. The faster the growth of new leaves the more likely a spiral pattern will occur. Apparently, not all of our sunflowers grew at the same rate. To be sure a more stringent test of this would be necessary before drawing any firm conclusions but it certainly suggests that more than genetic determinants are responsible for the growth patterns. Goodwin concludes from the research of Douady and Couder that whereas natural selection "may be involved in testing the stability of form it is in no sense a generator of biological form."[29] In the context of evolution he and Behe seem to be riding the same wave; he states with particular reference to the context of evolution that "there is an inherent rationality to life that makes it intelligible at a much deeper level than functional utility and historical accident."[30]

Historical accident . . . Thinking back a bit, it may seem strange in the light of this and all we have said above that science would confine evolution to the realm of historical accident and random mutation while the evolution of science has been anything but random. We have seen how the growth of science and the mentality of empirical thought have shared a cultural as well as a natural foundation. It was said above that science did not happen in a vacuum and we have seen how it developed in the wake of social trends that affected the world as a whole. Weaving in and out of this whole we have seen the intellectual strains of bygone ages alongside those of our modern period like the strains of a vast symphony. As we learn to hear these strains with eyes that see, we can begin to appreciate the many relationships that make all music, all life and all nature possible.

In the meantime, the orchestra plays on; the Chinese jar keeps turning. Among its many patterns we might recall the three gifts of Hermes Trismegistus and the attitude toward nature that cultivated a vision of how human beings are part of a greater whole. And if all this talk about qualities and relationships sounds like Aristotle's revenge, well, so be it. There was nothing keeping Aristotle from taking a dip in Heraclitus's river and coming out feeling, and looking, like a new man with that new Goethean look. After he dries himself off he might join us as we think back to how Maxwell spoke of the "mind's eye" with which Faraday saw a vision of holistic patterns regarding electro-magnetic fields. Till now such examples of the "mind's eye" in science have been largely dismissed by those who hold to the old Cartesian way of separating subject and object, human mind and nature, in a static framework of fragmented "realities."

Perhaps it is this "attitude" of separating man from nature that has led scientists to make of nature a lucky accident while they fill their cups of success and drink to the orderly march of science. Wait a minute! What was that I just wrote? Full cups? So that's what she meant! Remember our melodious caller? What she said about the jar being on empty because it was so full? Guess that goes to show that if you live in a Chinese jar, your own beginnings will come back to find you . . .

But before all our beginnings come to a happy end, we need to add a little more. We need to take note of how the on full on empty attitude is changing. Ways of improving our quality of thinking demand a view of how relationships, both natural and human, include the ever changing and interconnected patterns of existence. The need is for greater vision. As our wag once put it, "Baby, you've come a long way from the titanium view." And as evolution passes from reproduction to cultural and individual enlightenment, he might add that, "Baby, you've got a long way to go . . . " But the scientific vista promises in small but momentous ways to be a happier one. It introduces processes in our thinking that are inclusive rather exclusive and in so doing allows for inner growth to complement the natural processes in nature. Try that one on and we can wear that jar where the full and the empty are all the same size. And strum those patterns with a new kind of song that lives a new spin with a hum in the mind that sees where it ends with the trend on the mend . . . A little something to play upon as we take on the hard work to see this thing through . . .

To work our play and play our work we can pluck our luck by laying down the view with a few sober lines . . . lines that turn on the mind as we step out of the cave and discover how the nature of emergent order—otherwise known as the step from chaos—can be experienced in the attitude of play that leads to creative thinking. Goodwin speaks of "creation out of chaos" in this light:

> It is in playing and only in play that the individual child or adult is able to be creative and to use the whole personality, and it is only in being creative that the individual discovers the self.[31]

I only wonder if he has ever been to a Waldorf Kindergarten! Rudolf Steiner was keenly aware of the need for what he called creative play in the educational process. A growing mind needs rhythm and structure that breathe in and out of one another. Such breath is necessary for a mind in motion as feet and thought keep time with the feeling realm. When this happens the whole body responds in a way that allows it to become a vehicle and a recipient of a greater consciousness of a healthy and loving mind. In his *Philosophy of Freedom*, Steiner shows how the intuitive activity of thinking builds a vision of relationships that connect us with what we observe. I'm sure he would very cheerfully recognize a move toward the same awareness in the words Goodwin uses to close his book: "a science of qualities is necessarily a first person science that recognizes values as shared experiences, as states of participative awareness that link us to other organisms with bonds of sympathy, mutual recognition and respect." [32]

This is not to say that we can bid Descartes or the mechanistic paradigm good-bye. Old ideas die hard.[33] But a serious and viable alternative view has been made. It's a lead. One we must follow up on if we are to advance more consciously toward the more healing future.

And we step into the jar, take a deep breath, deeper than usual, and with a long pause we can see why the beginning and the end return . . .

Notes to Chapter 8

1. The concept of a dance rings especially true regarding Kekulé's dreams and the discovery of the benzene ring. In 1880, August von Stradonitz Kekulé, the father of structural chemistry, gave a speech during a celebration in his honor in which he stated that he owed his career to two visions. One, in 1854, occurred when he took a nap in a London omnibus and dreamed of a ring of carbon atoms dancing in a ring. The other occurred when he dozed off while sitting in front of a fireplace in Gant in 1861 and dreamed of a uroboros or serpent with its tail in its mouth. (*Histoire de la Chimie*, p. 199.) In Kekulé's mind these dreams confirmed ongoing research that coincided with discoveries he and others were making in the field of organic molecular structure. Is it a curious coincidence that the electrons in a benzene ring do indeed rotate about a ring of carbon atoms like the uroboros? Perhaps. But it bears noting that the resonance hybrid diagram of a benzene ring takes this electron movement into account by symbolizing the benzene ring with a circle inside a hexagon.

2. For research on the effect of electricity on healing and growth see the *Body Electric: Electromagnetism and the Foundation of Life*. The authors show that electricity affects cellular growth in bones all the way to the genetic level with what they call a "circuit of awareness."

3. I owe my interest in titanium to my dentist, Dr. Phillip McDonnell. It seems that titanium is used in dental reconstructions and other places where major

bone repair is needed because of this unique quality. Subsequent research into the matter has made the element a matter of increasing interest. It is pertinent here because it offers and excellent example of how a qualitative grouping of properties complements a holistic organic process.

4. Jar breaking is not to be confused with the paradigm shift made famous by Thomas Kuhn's *The Structure of Scientific Revolutions*. His description of paradigm shifts and how they are a natural part of science depicts a kind of changing of the guard as one set of authoritative theories replaces another. He writes that

> paradigms are not corrigible by normal (we would say "outside the jar") science at all. Instead, as we have already seen, normal science ultimately leads only to the recognition of anomalies and to crises. And these are terminated, not by deliberation and interpretation, but by a relatively sudden and unstructured event like the gestalt switch. Scientists then often speak of the "scales falling from the eyes" or of the "lightning flash" that "inundates" a previously obscure puzzle, enabling its components to be seen in a new way that for the first time permits a solution. (pp. 122–3)

Like Plato's prisoner getting out of his cave, in fact. In Kuhn's setting of the cave story, the prisoner goes back into the cave as something of an authority as he propounds another theory. Not any more, however. Jar breaking dispenses with paradigms and models altogether. Paradigms just aren't what they used to be—they are fast becoming lost in the sauce of the post-modern world, a world largely created by the rapid exchange of ideas fostered by the computer age. This connectedness of science enables a fast track *movement* of ideas that grows organically and as such is independent of the academic tendencies that allow ideas to get stuck in an authoritative rut. In this way Behe is perhaps more of a jar-spinner than a jar-breaker as the natural run of ideas gets things flowing. But he committed one cardinal sin according to the Order of the Jar: he didn't publish by process of peer review. This means that he didn't submit his views to an established scientific journal that submits its articles to a reviewing of established scientists in the field. We can guess from the reviews his book did get from some of these established peers that if he had done so the book would have never made it to the printing house. In fact, some seemed to imply in no uncertain scientific terms that it would barely make it to the outhouse. The main complaint is that he does not accept a thoroughly materialist explanation for complexity. As one reviewer states, "If our hypotheses about complexity are to be of any use, however, they will have to be materialist explanations grounded in material cause." (Robert Dorit, *American Scientist* online.) Another reviewer complains that any thing less brings science to a dead end. But it needs to be pointed out that materialism as a prerequisite for doing science is a philosophical attitude that has its roots in nineteenth century positivism. It is the basic tenet of positivism that any data not derived from experiment or formal logic cannot be regarded as valid: truth is that which can be objectively and experimentally verified. The problem with this statement, as R. G. Woolley points out, is that it cannot itself be experimentally verified. He writes that the proposition "fails the test it proclaims, i.e., in its own terms in cannot be true. It is quite evident that this formula for "truth" is beyond the limits of both experimental investigation (what experiment could be invoked to verify it?) and formal logic." (*Molecular Structure Conundrum*, p. 1082.) In other words, positivism and the materialist restriction that goes with it, are based not on experimental "truth" at all but on a philosophical attitude. The irony of this is that Behe does indeed subscribe to the positivist paradigm by basing his conclusions on experimentally established and demonstrable results, on what he calls an "open box." Quite the contrary, it was Darwin who based his theory on a "black box" of certain assumptions that could not be based on experiment because the science to do the experiments was lacking in his day. Behe and other proponents of complexity are pointing out how this in many ways has not changed and that neo-Darwinism, with its selfish genes (Richard Dawkins) and Anglo Saxon metaphor (survival of the fittest, genes in control, etc.) becomes more philosophy than science. It is precisely for this reason that Lynn Margulis calls neo-Darwinism a "minor twentieth century religious sect." (See note five below.) And what does this mean for us? Well, some good drama full of vital issues for one thing. And some rather spicy patterns on our jar. Keep turning.

5. *Darwin's Black Box*, p. 250

6. In a section entitled "The Natives Are Restless" Behe quotes Lynn Margulis'* comment on the matter (Lynn Margulis is Distinguished University Professor of Biology at the University of Massachussetts). She expresses the view that history will ultimately judge neo-Darwinism as "a minor twentieth-century religious sect within the sprawling religious persuasion of Anglo-Saxon biology." Ibid, p. 26.

7 *How the Leopard Changed Its Spots*, p. x

8. I'm not being as original here as it might sound. In fact, it was all done with mirrors. I was reflecting on your smile and Behe's mousetrap. He uses it as the quintessential example of an irreducible machine before he graduates to Rube Goldberg devices and the complexity of molecular systems. I just somehow connected the two. So perhaps I am being a bit original after all. At the very least it's just a little something you can think of the next time the camera man or woman tells you to say cheese. Or when you reflect upon the reflection of your original self in the mirror.

9. *How the Leopard Changed its Spots* , p. 136.

10. Ibid.

11. Not to say that Goodwin is alone in asking this question. In fact he makes it very clear that biology is rather alone in still adhering to the mechanical

model. Quantum mechanics has long since dispelled itself of the classical understanding of the word mechanical and has brought in such very non-mechancal understandings of nature as nonlocal causes, etc. (which Goodwin mentions). Scientists in other fields have long questioned and criticized the application of the mechanical model in biology and neo-Darwinism. To mention a few: Paul Weiss, Professor Emeritus of Rockefeller University *(System of Nature and Nature of Systems,* in *Toward a Man-Centered Medical Science,* pp. 17–63 and *Whither Life Science?* American Scientist, vol. 58, 1970); Loren Eisely *(Was Darwin Wrong About the Human Brain?* Harper's Magazine, 211:66–70, 1955) and of course the philosopher and physical chemist Michael Polanyi *(Life Transcending Physics and Chemistry,* C & N, August 21, 1967). Though Goodwin sounds new, the dates on these articles attest to the fact that his views echo a long standing wave in the recent history of science.

12. *How the Leopard Changed its Spots,* p, 197

13. Seven years after James Lovelock came out with his book *Gaia: A New Look at Life on Earth* in 1979, Lynn Margulis and Dorion Sagan co-authored *Microcosmos: Four Billion Years of Microbial Evolution.* Both books caught a new wave of thinking that looked at nature as a cooperative endeavor and offered fresh ideas to counter the competitive paradigm inherited from the nineteenth century (see note 14 below).

14. For more on the relation of music to holistic thinking, see Ronald Brady, "Goethe's Natural Science," in *Toward a Man-Centered Medical Science,* p. 157.

15. Lovelock, still ascribing to the mechanical model, likens the earth to a holistic cybernetic system (which he compares to an oven with a thermostat). Margulis also has a mechanistic view of how micro-organisms cooperate in mutual support systems to create complex organic machines. Be that as it may, both scientists made a major contribution by substituting a holistic paradigm for evolution that seriously countered the typical notion of survival of the fittest.

16. *The Continuing Tale of a Small Worm,* p.154.

17. Ibid

18. Ibid

19. *Caenorhabditis Elegans,* p. 41

20. *The Continuing Tale of a Small Worm,* p. 154

21. Ibid

22. Ibid.

23. *How the Leopard Changed Its Spots,* p. 181

24. Ibid, p. 140

25. Ibid, p. 182

26. The view from inside the jar is looking more and more like the view from outside Plato's cave. Leaving the cave is all about emergent order. The Pythagorean take of creation draws on ancient myths that tell how Chaos creates the Cosmos—which means the Perceived Universe and signifies Order. Inside the cave, in the darkness of the mind, there are only shadows, opinions and chaotic thought forms. Leaving this creates a condition of emergent order and the perceiving of the uni-verse as a whole. So in case you thought any of the above is new, well think again . . . (*Le Nombre d'Or*, vol. 2, p. 10)

    From inside the jar we can also see how we are a part of the turning world which is really more of a spiraling world where ideas and concepts return but are never quite the same. More and more it is this spiraling evolution of ideas that shapes the content of our lives and the identities we fashion from the connections we make. For this reason life is essentially nonlinear; it is circular. As we see how our identities depend on the relationships we make we become more in tune with the need to improve the quality of thinking that improves the quality of our interaction with a world that improves us as we improve it. And so the wheel turns, the jar spins and yet we see this from a still point inside ourselves.

27. I paraphrase Eliot's "still point of the turning world." (Burnt Norton—though the "still point" is a theme throughout the four parts of the poem.) And while was I getting a little help from the poem, you'd never guess what happened. Our friend with the melodious voice called and left her number on my answering service. Only it was in code—said it was contained in the last part of the poem. OK, I know that sounds hokey but metaphorically yours, it's the God's truth. Anyway, it's probably better if I leave it up to you to figure it out. If you do, give her a call, and find out for yourself where it's all going . . . Here's how the poem ends:

    > We shall not cease from exploration
    > And the end of all our exploring
    > Will be to arrive where we started
    > And know the place for the first time.
    > Through the unknown, remembered gate
    > When the last of earth left to discover
    > Is that which was the beginning;
    > At the source of the longest river
    > The voice of the hidden waterfall
    > And the children in the apple tree
    > Not known, because not looked for
    > But heard, half-heard, in the stillness
    > Between two waves of the sea.
    > Quick now, here, now, always—
    > A condition of complete simplicity

(Costing not less than everything)
And all shall be well and
All manner of thing shall be well
When the tongues of flame are in-folded
Into the crowned knot of fire
And the fire and the rose are one.

28. The experiment consists of dropping a ferrofluid (a liquid that can be magnetized) onto the center of a turning disc that is coated with a film of oil. The apparatus is subjected to a magnetic field that polarizes the drops as they fall, causing them to repel each other. As the drops fell onto the disc they are repelled equally from from each other to create equal angle (logarithmic) spirals that intersect at Fibonacci ratios like the spirals of a sunflower blossom. The patterns would change depending on the rate of drops. Goodwin used the patterns to demonstrate the phyllotaxes of leaves up a stem. (See *How the Leopard Changed Its Spots*, pp. 127–133)

29. Ibid, p. 133

30. Ibid, p. 116

31. Ibid, p. 201

32. Ibid, p. 237

33. One can generally rely on *Scientific American* [SA] or *American Scientist* to publish anything that advances the mechanical model. So it was not surprising to find in the October 2004 issue of SA another article on junk DNA and how it might provide the mechanism for evolutionary change. The article is written by John Mattick, an eminent name in the field of molecular biology, and describes the growing evidence that nonprotein encoding RNAs (introns) and other repetitive DNA material seem to have a regulatory function that controls the complexity of an organism. Suspicions that this might be the case grew when it was discovered that simpler life forms have more DNA than more complex life forms. Some amphibians have as much as five times the DNA as mammals and some amoebae have 1,000 times more! Mattick refers to our friend, c. elegans, in this light. C. elegans has 19,000 protein coding genes whereas humans have a mere 25,000. But it has also been discovered that the amount of nonprotein coding genes or so-called junk DNA in more complex animals is much higher and this has lead some scientists to suspect that the "junk" must have some function that contributes to the complexity of higher life forms. This has resulted in upgrading junk DNA with the name of "introns." Old ideas that introns were just so much junk are themselves getting junked as new ideas look for how introns create the ordering principle that controls complexity. In other words, in conformity to the mechanical model, introns are supposed to be the hand that turns the crank. Though the metaphor might seem a bit crude, it nevertheless describes the bottom

line. One of the basic tenets of the mechanical model is that organisms are controlled from the molecular level up. This presupposes an orderly pattern of cell growth. It comes therefore as no surprise to read where Mattick states the following as though he were reading right out of the guidebook for the mechanical model: "Throughout their evolution and development, organisms must navigate precise developmental pathways that are sensible and competitive, or else they die." ("The Hidden Genetic Program," John S. Mattick, *Scientific American*, vol. 291, nr. 4, October 2004, p. 66) And yet, as we have seen from the researchers who analyzed the cell growth of c. elegans, this is definitely not the case. There was nothing "sensible and competitive" not to mention "precise" in the way c. elegans grew, though it's arguable that the whole process achieved what might be termed sensible aims that made the organism a mature c. elegans. We of course have no way of knowing if c. elegans would consider itself "competitive." But more to the point, the case of c. elegans indicates very strongly that it was not the parts, be they DNA or introns, which achieved this "sensible" aim. Rather, it was the "outcome of a more holistic logic of molecular assembly." It was the organism functioning as a whole. We are told by the researchers who spent endless hours observing the growth of this little worm that its cells were "autonomous" and acting out of their individual history and not "as a result of interaction with other cells." And contrary to what Mattick seems to indicate, it didn't die. It grew up to be another faithful example of a full fledged c. elegans! In light of what Goodwin said above about how certain cultural norms slip into one's "scientific" thinking, the confusing context around the words "sensible" and "competitive" falls just short of being humorous. It might be irrelevant that John Mattick hails from Sidney, Australia. But it isn't irrelevant to recall that in accord with certain cultural ethics that trace at least some of their roots to our rich Anglo Saxon heritage, one must be "sensible and competitive" and, one might add, in control or else you die. Old ideas, it seems, do indeed die hard.

Bibliography

Allen, Paul M., ed. *A Christian Rosenkreutz Anthology*. Blauvelt, NY: Rudolf Steiner Publications, 1964.

Allen, Reginald E., ed. *Greek Philosophy: Thales to Aristotle*. New York: Macmillan Publishing, 1966.

Andrade, E. N. da C. *Rutherford and the Nature of the Atom*. New York: Doubleday, 1964.

Bauval, Robert & Adrian Gilbert, *The Orion Mystery: Unlocking the Secrets of the Pyramids*. New York: Crown Publishers, 1995.

Becker, Robert O. & Gary Selden. *The Body Electric: Electromagnetism and the Foundation of Life*. New York: Quill, 1985.

Behe, Michael. *Darwin's Black Box*, New York: Simon & Schuster, 1996.

Bensaude-Vincent, Berenadette & Isabelle Stengers, *Histoire de la Chimie*, Paris: Editions La Découverte & Syros, 2001.

Berthelot, M. *Les Origines de l'Achimie*. Paris, 1885.

Boorse, Henry A. & Lloyd Motz, ed.*The World of the Atom*, Basic Books, New York, 1976.

Brock, William A. *The Norton History of Chemistry*. W. W. Norton & Company, New York, 1993.

Cavalieri, Liebe F. *Double Edged Helix: Genetic Engineering in the Real World*. New York: Preager Scientific, 1985.

Curie, Pierre and Marie Curie. "On a New Radioactive Substance Contained in Pitchblende," *Comptes Rendus*, 127 (1898), pp. 175–178.

Eliot, T.S. *Complete Poems, 1909–1962*. New York: Harcourt Brace World, 1952.

Fauque, Danielle. *Lavoisier: La Naissance de la Chimie Moderne*, Paris: Editions Vuibert, 2003.

Finley, Alexander. *A Hundred Years of Chemistry*. London: Duckworth & Co., 1965.

Goodwin, Brian. *How the Leopard Changed its Spots: The Evolution of Complexity*. New York: Charles Scribner's Sons, 1994.

Heisenberg, Werner. *Physics and Philosophy: The Revolution in Modern Science*. New York: Harper & Row, 1958.

Jacq, Christian. *Le Message Initiatique Des Cathédrals*. La Maison de Vie, 2003.

Jaffe, Bernard. *Crucibles: The Story of Chemistry*. New York: Dover Publications, 1976.

Kuhn, Thomas S. *The Structure of Scientific Revolutions*, 2nd ed. Chicago: University of Chicago Press, 1970.

Lehrs, Ernst. *Spiritual Science, Electricity and Michael Faraday*. London: Rudolf Steiner Press, 1975.

Lewin, R. "The Continuing Tale of a Small Worm," *Science*, vol. 225, July 13, 1984.

Lucretius, *On the Nature of Things*, Martin Ferguson Smith, ed. Cambridge: Hackett, 1969.

Luft, Robert. *Dictionaire des Corps Purs Simples de la Chimie*. Nantes: Cultures et Techniques, 1997.

Margulis, Lynn & Dorion Sagan, *Microcosmos: Four Billion Years of Microbial Evolution*. New York: Simon & Schuster, Inc., 1986.

Marx, J. L. "Caenorhabditis elegans: Getting to Know You," *Science*, vol. 225, July 6, 1984.

Mattick, John S. "The Hidden Genetic Program of Complex Organisms," *Scientific American*, vol. 291, nr. 4, October 2004.

Pauling, Linus. *General Chemistry*. New York: Dover Publications, 1988.

Pierce, John R. *The Science of Musical Sound*. New York: Scientific American Books, 1983.

Poirier, Jean-Pierre. *Lavoisier: Chemist, Biologist, Economist*. Philadelphia: University of Pennsylvania Press, 1996.

Plutarch, *Moralia*, vol, V. Cambridge, MA: Harvard UP 1993.

Paracelsus, *Selected Writings*, Bollingen Series XXVIII, Princeton: Princeton UP, 1969.

Ross, Sir David. *Aristotle*. New York: University Paperbacks, 1964.

Schaefer, Karl E., Herbert Hensel & Ronald Brady, eds. *Toward a Man-Centered Medical Science*. Mount Kisco, NY: Futura Publishing, 1977.

Solé, Ricard & Brian Goodwin. *Signs of Life*. New York: Basic Books, 2000.

Steiner, Rudolf. *The Philosophy of Freedom*. London: Rudolf Steiner Press, 1964.

Steiner, Rudolf. *The Warmth Course*. Spring Valley, NY: Mercury Press, 1980.

Thomson, J. J. "Cathode Rays," *Philosophical Magazine*, 44 (1897), pp. 293–311.

Tompkins, Peter. *Secrets of the Great Pyramid*. New York: Harper & Row, 1971.

Veatch, Henry B. *Aristotle: A Contemporary Appreciation*. Bloomington, IN: Indiana UP, 1974.

Wellman, Kathleen. *La Metrie, Medicine, Philosophy and Enlightenment*. Durham, NC: Duke University Press, 1992.

Woolley, R. G. "The Molecular Structure Conundrum," *Journal of Chemical Education*, vol. 62, (1985).